Smarter Together!
Collaboration and Equity in the Elementary Math Classroom

Helen Featherstone
Michigan State University, East Lansing, Michigan

Sandra Crespo
Michigan State University, East Lansing, Michigan

Lisa M. Jilk
University of Washington, Seattle, Washington

Joy A. Oslund
Alma College, Alma, Michigan

Amy Noelle Parks
University of Georgia, Athens, Georgia

Marcy B. Wood
University of Arizona, Tucson, Arizona

NATIONAL COUNCIL OF
TEACHERS OF MATHEMATICS

Copyright © 2011 by
The National Council of Teachers of Mathematics, Inc.
1906 Association Drive, Reston, VA 20191-1502
(703) 620-9840; (800) 235-7566; www.nctm.org
All rights reserved

Library of Congress Cataloging-in-Publication Data
Smarter together! Collaboration and equity in elementary mathematics /
Helen Featherstone ... [et al.].
    p. cm.
  ISBN 978-0-87353-656-1
  1. Mathematics--Study and teaching (Elementary) 2. Team learning
approach in education. I. Featherstone, Helen, 1944-
  QA20.G76S63 2011
  372.7'044--dc23
                              2011030530

The National Council of Teachers of Mathematics is a public voice of
mathematics education, supporting teachers to ensure equitable mathematics
learning of the highest quality for all students through vision, leadership,
professional development, and research.

Printed in the United States of America

Table of Contents

Preface

Originally, all of us were teachers in public schools. Most of us taught in elementary schools, and we struggled to teach math in ways that would enable all our students to build robust understandings of the elementary school math curriculum. Most of our students did well, enjoying the chance to move beyond computation and think about the reasons why the rules they had learned might work. But we were never satisfied that *all* our students were learning well. Some children always seemed to hang back, unable or unwilling to join group investigations, struggling even to begin work on the problems that their classmates were solving. Others did not seem to be challenged enough. So, not surprisingly, when we came to Michigan State University and began working with other teachers and prospective teachers in the certification program, we worried about giving our students the tools they needed to succeed in teaching powerful mathematics to all *their* students, a goal dear to their hearts and to ours.

Then we learned about complex instruction, the insights and strategies that were enabling a scattering of other U. S. elementary and secondary school teachers to address the problems that make it so difficult to engage the full range of children in our classrooms with challenging, important mathematics. Complex instruction seemed to enable teachers—even prospective ones—to engage children who had formerly remained silent and uninvolved during serious mathematical inquiry. Finally we could offer help to teachers who wanted to know how to engage "my low kids."

We address this book to elementary school teachers and interested teacher educators who want to teach serious mathematics to all their students, to challenge children who now finish their work long before their classmates, and to involve more fully children who currently try to remain invisible in math class and show little understanding of the math we try to teach them. The book is about using complex instruction to teach elementary school mathematics. It builds on work that the six authors—Helen Featherstone, Sandra Crespo, Lisa Jilk, Joy Oslund, Amy Parks, and Marcy Wood—did together at Michigan State University to teach teachers and prospective teachers in our elementary school teacher preparation program. We strove to make complex instruction a part of their mathematics teaching practice, with the goal of teaching challenging mathematics conceptually to *all* children in their increasingly diverse classrooms.

But our work with elementary school children, teachers, and prospective teachers did not start in 2006, when we first made complex instruction a part of elementary math education at Michigan State University. It started years earlier with the work of one of us, Lisa Jilk, as a high school math teacher, first in Minnesota and later in California. In the prologue, Lisa tells the story of her development as a high school math teacher, of the work that brought her to believe strongly that complex instruction could help her and others teach challenging mathematics conceptually to students in highly diverse classrooms and see all these young people learn.

This book is the result of collaboration among the six authors, to which all contributed equally, though differently. The order of names on the title page reflects alphabetical conventions and not a judgment on anyone's relative importance. Featherstone is listed first because she wrote the first draft of many, though by no means all, chapters. All six authors, however, discussed each chapter both before and after each of many drafts. All six identified issues that needed to be addressed and supplied new language. All contributed vitally important ideas and experiences, both through chapter revisions and through our discussions of the text. All shaped the manuscript significantly. Without the contributions of any one of us, the book would be different and not as good.

—Helen Featherstone, Sandra Crespo, Lisa Jilk,
Joy Oslund, Amy Parks, and Marcy Wood
September 2011

Prologue

This Felt Right to Me

Lisa Jilk

I began my teaching career believing there was a better way to learn math than the one I had experienced as a student. I determined that I would figure out how to create a classroom that matched this belief. Although I enjoyed school mathematics, I knew it only as a class where I quietly sat at my desk, copied the teachers' notes from the chalkboard into my notebook, and practiced the same problems repeatedly until I could recognize the problem type and reproduce the procedures that most often led me to a correct answer. To this day, I don't know why I continued to enroll in math courses long after I completed both high school and college graduation requirements. Maybe on some level I felt successful. I earned good grades, and I enjoyed following rules. Although I rarely understood where the rules came from or why they worked, choosing formulas, plugging in numbers, and following procedures gave me a sense of control that I often lacked in other parts of my life.

I struggled through my first three years of teaching math in a comprehensive high school in a first-ring suburb of Minneapolis. I touted teaching for understanding and collaborative learning as a preservice teacher, but I was completely ill-equipped to turn my vision into a reality. I sought out programs that afforded alternative teaching and learning practices, but I failed miserably with a toolkit that included a traditional math curriculum, some lofty theoretical articles from my preservice program, and a handful of kind, very well-intentioned colleagues. Although my classroom was a place where young people felt comfortable, respected, smart, and safe, I am not confident that they ever learned enough important mathematics.

In my third year of teaching, the school asked me to teach a third track of students who were not finding success in our math program. The department had decided to adopt the Interactive Math Program (IMP), a four-year program of problem-based mathematics supported by the National Science Foundation, with hopes that an alternative, more problem-based curriculum would serve these young people better. Teaching IMP gave me professional development opportunities that helped me shape new opinions about myself as a math learner. In school and in college I had, more often than not, entered math classrooms shy and fearful, with the goal of sitting quietly unnoticed until I could go home and try to make sense of the notes I took in class. Sitting in the back of the room and not participating was not an option in IMP workshops. The program positioned us, the teachers, as learners, put us in small groups, and asked us to investigate, probe, and question the math—exactly what we would eventually ask our own students to do. Experiencing the fear of not knowing, but working through my confusion with the support of the facilitators' good questions and scaffolding strategies, gave me the confidence to say, "I don't know, but I'll try to figure it out." For the first time, I was really learning math. I was struggling with ideas, drawing pictures, having conversations, and exploring concepts, and no one ever told me I was wrong or stupid. Instead, they encouraged my ideas and applauded my perseverance. This was new and exciting, and in many ways this experience brought me out of the mathematical closet. I was very proud of my newly discovered abilities, and I finally understood how my participation in my own learning added to my mathematical competence. I knew then what I needed to do for kids.

I left Minneapolis and moved to northern California in the late 1990s. There, I found Railside High School (a pseudonym), one of the high schools that participated in a five-year, longitudinal research study led by Jo Boaler of Stanford University (Boaler and Staples 2008). Joanne Lieberman, currently a professor of mathematics education at California State University—Monterey Bay, was then teaching a course in my graduate program. She had recently completed her dissertation about the Railside math department as a professional learning community, and she suggested that I check the school out. I clearly remember the day of that visit, because my experience exemplified exactly what the Railside math program was all about. (Although I use the past tense when I describe the collaboration and teaching that took place at Railside when I was a teacher there, this same kind of work still happens at Railside today.) Hoping not to disturb anyone, I found a seat in the back corner of an algebra class. Little did I know that I had absolutely no chance of being overlooked or left to sit quietly and observe. Carlos Cabana, the department chair at the time, asked me up to the front of the room to help launch a lesson that I knew absolutely nothing about. I took a seat at the overhead and recorded information generated from a whole-class discussion. There I was, a stranger who just happened to stop by, gradually being drawn into this classroom community, and no one blinked an eye. Conversation moved easily among Carlos, the students, and me, and I felt connected to these kids and immediately invested in what they were learning. I soon learned that this way of being in math classrooms was commonplace at Railside. No one was a bystander, and everyone had something to offer and something to learn.

Soon after that memorable visit, Railside hired me as a long-term substitute teacher. I began my full-time tenure the following fall. When I joined the Railside math team in 1997, the department already had a strong reputation for collaborating in grade-level teams, creating curriculum around big mathematical ideas, and using groupwork as its primary mode of teaching and learning in all courses. I was very attracted to this model, and I was thrilled when Railside hired me. I imagined Railside as a place where I might learn to actualize the vision of teaching and learning math that had for years felt so elusive. Prior to my arrival at Railside, the department had detracked its math courses and begun a partnership with Elizabeth Cohen and Rachel Lotan at Stanford University in an effort to increase achievement and meet the demands of working with heterogeneous classes. The department had eliminated all prealgebra classes, and it now randomly placed every incoming, English-speaking ninth grader in an algebra class. It also offered a Sheltered Algebra course for students who were English language learners. Having eliminated ability tracking for tenth, eleventh, and twelfth graders as well, the department integrated most students with special needs into mainstream classes and moved all students through a college-preparatory program that included two years of algebra, geometry, trigonometry, precalculus, and calculus.

The student population at Railside has remained quite consistent over the years. In the late 1990s approximately one-third of the students were Latino, one quarter each European American and African American, and seventeen percent Asian American and Pacific Islanders. Ten percent were English language learners; 24 percent qualified for free or reduced-price lunch. At the time, Cohen and Lotan had developed the Program for Complex Instruction based on research focused on learning how to teach, at a uniformly high level, students who have a diverse range of prior knowledge, language abilities, and social status. Cohen and Lotan's work assumes that every child is academically competent. That work takes up the serious issues of social and academic status that run rampant in our society and often impede students' participation and learning. In our world, people often make decisions about other people's intellectual abilities on the basis of certain characteristics that the community values. Often, for example, because most communities have assigned more value to English than to other languages, people assume that others who do not speak English fluently do not have the same competence as those who are native English speakers. In the same way, many Americans believe that people with white skin are more intellectually capable than people with black or brown skin.

People then act from these assumptions. A student might ignore a Punjabi classmate's ideas and dominate the small-group presentation because she assumes that her partner's lack of fluency in English makes him dumb at math. A young black man may actively disengage from collaboratively doing a math task because he believes that the white students in his small group have more math skills. In both these examples, the girl who dominates or the young man who disengages, students are participating unequally and therefore missing out on valuable learning. Seeing how this might happen for the young black man is easy: no participation usually leads to less learning. However, what is more difficult to understand, but is just as true, is how the person who dominates the conversation, who takes over the task and does all the work, is also not learning enough mathematics. If we believe that learning mathematics requires participation with others around rigorous content, then the student who silences his groupmates will inevitably learn less than the student who engages with them in discussing the math.

Our students arrive in our classrooms with most of the assumptions about competence that pervade U.S. culture, and then they act on them. Some students do not want to participate in our classes because they do not feel smart. As I have said, that was me for many years. Some students completely dominate activities and conversations because they feel more intellectually able than their peers. We all know these young people. We have all seen these behaviors and chalked them up to personality, something we believe ourselves powerless to change. However, what we believe about our own intellectual abilities, and those of others, is very often a social construction reinforced over time rather than a mirror of reality.

At Railside, we math teachers held the unwavering belief that *all* students are smart and *all* students can succeed in learning mathematics. Therefore, Railside's partnership with Cohen and Lotan was an opportunity to marry beliefs with practice. When a school detracks courses, the vast range of prior math experiences and understandings that exist in one classroom can overwhelm even the best teachers. In addition, this diversity can make academic and social status issues more visible. Cohen and Lotan's Program for Complex Instruction gave Railside's math teachers real tools for addressing these challenges. It helped us create classrooms where students worked together to dig in and make sense of rigorous mathematics, where students took responsibility for their learning and that of others, and where young people with diverse schooling and home experiences could work together to achieve their academic goals.

This felt right to me. This felt normal. A stance toward teaching and learning that prioritized the collective rather than the individual and collaboration rather than competition coincided with much of how I tried to live my life outside school. Learning is not a zero-sum game. Everyone can learn. Success does not have to come at another person's expense, and working together only makes us all stronger, smarter, and happier in the long run. My experience as a teacher at Railside supported these beliefs, and I watched as our students came to believe the same.

I remember the first Sheltered Algebra class that I taught at Railside. Thirty-two students ranging from 14 to 18 years of age, from different countries around the world, came to me with a hugely diverse set of previous math experiences, skills, and conceptual understandings. Most of these young people were from Mexico, Nicaragua, and El Salvador; several had recently arrived from Vietnam, China, and India. My responsibility for ensuring their success in algebra completely overwhelmed me. None of them were yet fluent in English, and I had not learned how to speak Spanish, much less Vietnamese or Punjabi. I learned that more than half of them had not successfully completed their eighth-grade math course.

I knew immediately that I absolutely could not do this work alone, and I was certain that the students looked at me on that first day and thought the very same thing. They probably wondered, "How in the world is *that white* woman going to teach us?" The very fact that I could not meet all my students' needs required me to rely on them for help. I needed their help to translate directions, to lean across tables and share ideas, to come up to the overhead and explain their reasons. I needed Chris to sit with Billy, because Billy was still angry with his parents for bringing him to California from Vietnam. I needed Lupe to tell Norberto and Ricky to sit down and be quiet every five minutes, because Lupe had "mothered" these young men throughout middle school and had an amazing skill for holding them accountable for doing their work. I needed Catherine to show Doris what it meant to combine like terms, because I didn't know how to say "combine like terms" in Shanghainese. Pablo had failed every math class he had ever taken, but he was 17 years old and needed to graduate. It turned out that I needed Pablo to show his group how to use rubber bands to create geometric shapes on a geoboard so they could apply their area formulas. What we in that classroom needed each day for mathematical success did not come from any *one* of us. It came from *all* of us using our strengths and skills to help everyone get the math they needed to learn.

This commitment to learning together, holding one another accountable for everyone's learning, was also being played out in the school's math department. Being a member of the Railside mathematics department meant that I was part of a team of teachers deeply committed to ensuring that every student who walked through our classroom doors learned rich, powerful mathematics. No teacher in the department blamed the kids, the parents, or the community for difficulties they encountered trying to make this happen. Of course, we all wished for more money, time, and resources. Ultimately, though, figuring out how to engage each student in learning mathematics, and how to support their academic success, was our job. Starting with this commitment meant that we needed one another as colleagues. No one of us could take on this tremendous challenge alone. We met together weekly to create tasks and solve the never-ending list of problems and challenges we faced in our classrooms. We discussed students' understanding and misconceptions, crafted probing questions, analyzed students' work, and discussed our observations of one another's classrooms. We worked with administration and counselors to streamline schedules and figure out how to support our students. We launched each year with Algebra Week, a weeklong teachers' working retreat where we reconnected after summer vacation, reflected on the previous year's successes and challenges, and constructed new lessons and units for our ever-changing curriculum.

We worked tremendously hard to cocreate a mathematics program along with safe, positive communities where students recognized their own mathematical competences and learned to work together. This kind of commitment and support for one another helped English language learners successfully complete more upper-level math courses than I had ever seen before. I often heard students proudly state that they were smart in math and that math was their favorite subject. I watched 30 percent of 2002's graduating class successfully complete Advanced Placement Calculus. I learned that gender, race, language, or class do not determine a young person's ability to learn mathematics. Railside High's students and math department forever changed my life. They confirmed my beliefs that anyone can learn math and everyone should have the resources and support necessary to do so.

Take Carrie, a junior in my class, who was taking her first regular algebra class outside the special education department. Carrie came to me feeling stupid. She was terrified of being in this algebra class, because she was so comfortable with the special education teachers and program. Although I worked hard to convince Carrie that she could be successful with us, she did not finally buy into that idea until she started working with a group of young women from Mexico.

Carrie, Claudia, Maria, and Jensy were working together to build a "monster dog" that was six times the size of the original dog I had given them and to find the new dog's volume and surface area. For a group of four ninth graders who did not all speak the same language, this was a big project. Our classroom norms required that teams stay together on a task. No one was supposed to move ahead or go faster than the rest of the group, and we did not allow teams to split the work up into pieces for individuals to do. Together, the young women needed to figure out how large the new dog should be, build it out of graph paper, and write a final report to explain and justify their solution.

After a day of struggling with this task, Carrie wanted to talk to me. She looked concerned and asked me if this was a "special" class. I wasn't sure what she meant, and I was cautious about labeling this class Sheltered Algebra. Given Carrie's previous experiences in special-education math classes, I wanted her to feel good about her placement in *this* algebra course. I wanted her to feel comfortable and be willing to take risks with her peers. I wanted other students to recognize Carrie's good math ideas, and I wanted her to participate more often. I knew that Carrie could be successful in this course, and I wanted her to believe this, too. When I pressed Carrie to tell me what she was worried about, she said, "I feel so smart in here! I get to speak Spanish all the time, because they need my help, and I'm really learning a lot of math." I was completely shocked. This young woman, who lacked academic confidence and who the school had positioned as a less-than-stellar special-education student for many years, was worried that something was wrong because she suddenly felt so smart!

I learned that day that Carrie was fluent in both Spanish and Portuguese. A math classroom where students worked individually and rarely spoke to each other might never have valued such language expertise. Carrie, however, found that this class desperately needed her strengths in Spanish, not because people wanted to learn Spanish, but because students who spoke Spanish needed access to math. Further, Carrie wasn't just a translator. In this classroom, she became a math learner who used both Spanish and English to interact with others while she and they learned algebra together. Carrie's Spanish fluency gave the Spanish-speaking students access to the math; simultaneously, Carrie gained access to new math ideas when she worked collaboratively with her team using Spanish. The nature of the Monster Dog task, the norms for collaboration, and the focus on students' strengths enabled these young women to complete their task successfully and to learn important mathematics.

When I left Railside to pursue graduate studies, I imagined that I would again join a community of like-minded educators committed to learning more about how to develop high-quality math instruction in our schools. Although most of those I met in the university wanted the best for all students, I realize now how naïve I was to think that we all had the same vision of how this should look. I struggled as I tried to share my Railside experiences and adapt these ideas to preservice teacher education. I learned a lot in graduate school. One thing that has stuck with me is that timing is crucial: no matter how good an idea is, until it meets opportunity, it often goes nowhere. After three years of reading, writing, talking, and exploring how my world of complex instruction might fit into the university, I saw things begin to happen. A few of us former Railside math teachers had begun sharing our work around the country. Then Helen Featherstone and Marcy Wood approached me. First, they wanted us to teach complex instruction to the prospective elementary school teachers in their classes. Then they wanted us to collaborate with other elementary math education instructors at the university, instructors committed to helping preservice teachers learn more about instructional practices that supported equitable participation in mathematical sense making. Bam! Our worlds collided. We all had things to learn with and from one another, and we used this opportunity to help preservice teachers believe in their own mathematical abilities and their capacity to achieve more equitable outcomes with students.

I am especially grateful for the chance to think and work with this very thoughtful, talented group of women. They challenge me to think about complex instruction in new ways, and they force me to clarify the connections among my beliefs and practices. This group has moved this work forward in ways important for math education in general and preservice elementary teachers' work in particular.

Complex instruction is not a magic pill. It contains no formula or checklist to follow. In addition to strong content knowledge, knowledge about students, and knowledge about how students think and learn math, successfully implementing complex instruction requires a deep belief that all students can learn, that teaching is more than delivering information, and that learning is more than listening. Complex instruction is, in fact, complex. Its components intertwine, and there is always too much to attend to. Given some extra time, resources, and support, very committed groups of math teachers at Railside—and now elsewhere—are using these ideas and practices, and they are making it work! Yes, *we* can!

This book offers a starting point for teachers who believe as I did early in my teaching career. I held great hopes for my students, but challenges that seemed impossibly immense overwhelmed me. This book offers a place to start growing new practices for helping students both make sense of mathematics and see mathematics as a social activity, as something worth doing together. My wish for those who read this book is that they trust their students and themselves to create the kinds of classrooms that inspire everyone to engage and learn.

Reference

Boaler, Jo, and Megan Staples. "Creating Mathematical Futures through an Equitable Teaching Approach: The Case of Railside School." *Teachers College Record* 110, no. 3 (March 2008): 608–45.

Introduction

The children in the fourth-grade class that Elise Murray has been teaching since August have been working on multiplication for several weeks. On this cold February morning, however, Elise has decided to do something quite different with the math lesson: she divides the children into groups of four and tells the groups to work together on a problem that requires three–dimensional spatial reasoning. Each group has a bag of painted wooden cubes, a set of clues, and an assignment card that tells them to build a structure that satisfies all the clues. Elise keeps an eye on the groups and takes notes on a clipboard while the groups are working. After about 15 minutes, she notes that the three boys in the group by the window have, after considerable discussion, created an arrangement that satisfies some, but not all, of the clues. Hoping to get the children to identify the discrepancy, Elise asks them to read all the clues aloud and check their structure to see if it fulfills the conditions that each clue sets. Forced to recognize their solution's inadequacy, the three boys, all confident math students, stare first at the structure and then at the clues.

Annette, the only girl in the group, moved into the school district only a month earlier and is struggling academically and socially. She is not confident about her math skills, and the class seldom hears her voice during math lessons. On this day, she has watched the other children silently. Elise, realizing that the boys are ignoring their quiet groupmate, positions herself at the other end of the table from Annette. Elise wants to ensure that the boys will hear anything the girl says to her teacher. As Elise kneels next to the table, urging the group to consider *all* the clue cards, she hears Annette whisper softly, "The green block has to go in the middle." This is the insight that the group needs in order to create a configuration that satisfies all the clues, but it goes unnoticed as the three boys speak confidently about their ideas. Leaning forward as though to hear Annette's contribution more clearly, Elise asks the girl to repeat her comment, speaking louder. Again Annette asserts, a little more audibly, "The green block has to go in the middle."

This time, after surveying the blocks and clues, one of the boys responds, "That's so smart! That's so smart! That's what we should do!"

Glancing across the table, Elise sees a broad smile spreading across Annette's face.

Over the past four decades, many educators and psychologists have argued that children and adolescents can learn a great deal from working in small groups (see, for example, Frey, Fisher, and Everlove [2009]; Janssen et al. [2010]). In addition to the obvious opportunities to learn and practice social skills like listening, questioning others, and cooperation, these educators contend that the extended opportunity to articulate ideas, work on making sense of the ideas that others struggle to express, and frame new ideas based on groupmates' multiple perspectives leads to conceptual development and cognitive growth. Indeed, Ben-Ari and Kedem-Friedrich (2000), of Ben-Ilan University in Israel, have data showing that the amount of verbal interaction in a small group associates significantly with learning: the more a child talks, the more she learns. This should be no surprise: we know from the work of psychologists Jean Piaget (1959) and Lev Vygotsky (1978), and from Anna Sfard's (2008) more recent work, that learning and thinking are social processes. We learn the skills that make us who we are from observing, listening, and experimenting in social contexts. Learning is a kind of cognitive apprenticeship. Lave and Wenger (1991) and Brown, Collins, and Duguid (1989) have shown that novices learn skills like tailoring and processing insurance claims by participating, marginally and later more centrally, in a group that is doing this work.

Although many educators have argued for groupwork's benefits, both social and cognitive, what actually takes place in the groups is often very different from what we hope when we recommend it. When we, the authors, as teacher educators, assign our students readings on the benefits of groupwork, some of the prospective teachers in our classes look skeptical. When pressed, they report that their own experiences of working with other students in small groups were mixed. Although sometimes groupwork was fun, participation was often very uneven. They recall occasions when a few students did all the work while others chatted and giggled with friends. We can't disagree. We have seen these same things happen in our own classrooms—elementary and secondary school and university—and in those of our students and colleagues. Some teachers report similar frustrations when working in groups with colleagues in professional development settings.

Complex instruction is a response to the paradox of groupwork, to the fact that groupwork offers so much potential but often deteriorates into a situation in which only a few students seem to learn. Complex instruction is a set of ideas and strategies for addressing the problems that seem so often to bedevil groupwork and block powerful learning for children. This book provides guidance to readers in using these strategies and ideas—in using complex instruction—to teach challenging mathematics to all children in the twenty-first century's heterogeneous classrooms.

The late Elizabeth Cohen of Stanford University coined the term *complex instruction*. It thus makes sense to start with the definition that she used in her work (Cohen et al. 2002, p. 1047):

> □ Complex instruction is a set of strategies for creating equitable classrooms. Using these strategies, teachers can teach to a high intellectual level in academically and linguistically
> □ heterogeneous classrooms.

The name *complex instruction* is intimidating; we probably would not have chosen it ourselves, since we are acutely aware that many teachers, on hearing the name, back away, at least metaphorically, thinking that regular teaching is quite complex enough. However, we have come to think that the term *complex instruction* expresses something important about learning mathematics.

One could describe the paradigm that has dominated Western thinking about teaching and learning for millennia as a linear one, and this is particularly true of mathematics. This paradigm identifies the teacher's job as deciding what the lesson will teach, dividing this material into manageable chunks, organizing these chunks logically, and then walking learners through them in a logical order. In mathematics, we see this as a ladder. Each skill builds on the one before, and if we break them down appropriately and present them in a logical order, none should confuse the learner.

In fact, however, mathematics is not a ladder, but a web, or, if we prefer a playground analogy, a jungle gym. Almost always, many ways exist to solve a math problem, even a simple one. These different ways build on different understandings and different ways of making sense of new mathematical ideas. Most people are happy to settle for one way of understanding an idea. Once they have found an answer, they move on to the next problem—or something else, perhaps recess. Complex instruction derives from (1) a different understanding of how people learn, and (2) a different image of what it means to understand a mathematical idea or even an algorithm. It builds on the idea that learning is complex, and that learners, because they bring different ideas and understandings to a novel problem, will make sense of the learning challenge it presents in multiple ways.

When a group of children are working on a problem together, the very complexity of the network of ideas they bring to this problem serves as a resource to a group's members. This resource's value derives first from the nature of learning generally, from the fact that, because learners come to a new idea with different background knowledge, they will find different paths into the problem accessible. For some children, visualizing 15×49 as 49 boxes of cookies with 15 cookies in each box works; for others, a rectangle of 15 by 49 unit squares is much more helpful. But the diversity of ideas about a problem is a resource for another reason, one deriving from the nature of mathematics: mathematics is a network of ideas related to one another in many different ways. Working with a group that identifies more than one way to solve, or work on, a problem makes this important understanding of the discipline visible.

When Cohen watched children working in groups in elementary school classrooms, she saw what the prospective teachers that we teach have seen: some children took charge and told others what to do, whereas a few said, and often did, next to nothing. But as a sociologist, Cohen brought a somewhat unusual lens to her observations of these classrooms. When she watched children working in small groups, she saw differences in status, rather than in ability or motivation, shaping who talked, who others listened to, and whose ideas directed what the group decided to do with the assignment. We will say much more about status—what it is, and how it affects classroom interactions and learning—in chapter 2. Here, we will simply offer Cohen's (1994, p. 27) working definition of a status ordering, "… an agreed upon social ranking where everyone feels it is better to have a high rank within the status order than a low rank."

Where teachers and other researchers saw children sitting passively, apparently unwilling to contribute even minimally to the group's work, Cohen often saw children whom their higher-status classmates had shut out—sometimes subtly, sometimes not. High-status children often excluded others from group deliberations. Even when their lower-status classmates did offer suggestions, as Elise's student, Annette, did, the higher-status students ignored their ideas.

Cohen also noted that children were frequently working in groups on problems that most of them could have solved fairly readily alone. These tasks were not particularly likely to generate joint deliberation and inquiry. In fact, when observing in elementary school classrooms, we often see groups dividing up the labor (i.e., "Milo can do the first two problems, Carlos can do 3 and 4, and Denise gets 5 and 6, and I'll do the last two"), so that the group will finish quickly, with minimum effort for all members. This way children avoid the drudgery of doing multiple computations; they also avoid the potential intellectual benefits of conversation.

To address the problems that she uncovered in the classrooms, Cohen worked with colleagues at Stanford and San Francisco–area teachers to find ways to minimize the effects of status on students' participation. She describes insights and strategies that resulted from these collaborations in *Designing Groupwork: Strategies for the Heterogeneous Classroom* (Cohen 1994). These ideas will aid immensely any teachers who use groupwork in their classrooms, no matter what subjects they teach.

So, what might a classroom look like when teachers and students are "doing complex instruction?" On the surface, it would probably look much like any classroom in which children are working in groups on an academic assignment. However, in complex instruction, certain features listed below are in place.

- Children are working in groups of three to five on a "groupworthy" task. We will explore what makes a task groupworthy much more fully in chapters 4 and 7. What we want to say here is that a task is groupworthy if
  — it challenges *all* children in the class intellectually;
  — children can approach it fruitfully in several different ways; and
  — it engages students with important ideas in one or more academic disciplines.

- The teacher has composed the groups randomly. Their membership changes regularly, usually every couple of weeks.

- The teacher structures the assignment so that *all* children must participate in the *intellectual* work of problem solving (see chapters 5 and 6).

- The teacher holds all children in each group responsible for understanding the work that their group has done together. This means, for example, that every student must be prepared to report to the rest of the class on his or her group's problem-solving strategy and solution.

- Each child has a defined role to play in his or her group. The roles distribute both the group's practical work and its intellectual work (see chapter 3).

- Teachers attend to potential problems relating to participation and status and have strategies for addressing these problems if they observe them (see chapter 6).

If we look again at Elise Murray's math lesson above, we can see some of these features of complex instruction. To begin with, the task that the children are working on appears to challenge all of them without requiring that they apply rules or procedures that only some of the children know. In addition, this problem engages the children with an essential challenge of mathematical reasoning: it requires them to hold multiple conditions in their heads simultaneously, rather than deal with each in sequence; and it carries them into three-dimensional geometry. The teacher is clearly attending to issues of participation. She uses several strategies designed to increase the participation of the low-status student she is addressing and to ensure that the high-status students notice and appreciate the intellectual contribution of the classmate they have ignored.

Elise's account of this lesson helps us see the ways in which status differences in a randomly composed group of three socially and academically successful boys and one low-status girl were restricting all four children's learning opportunities. Annette was losing the chance to talk to others about an interesting spatial-reasoning task. Similarly, the boys were unwittingly denying themselves access to an idea that could have moved them beyond their own restricted problem-solving strategies. Elise modeled one approach to addressing this status issue when she positioned herself so that the typically soft-spoken Annette had to raise her voice in order to make herself audible to her teacher. This strategy increased the likelihood that the boys would actually notice Annette's contribution and realize that the classmate they were ignoring had valuable ideas. Elise's lesson also demonstrates the way in which a different kind of problem can both (1) extend the thinking of children who, because they have mastered many arithmetic skills, have lost the habit of actually *thinking* about math problems; and (2) illuminate the intellectual strengths of a child who is currently better at logical and spatial problems than at computations.

Since Plato's time, people have spoken and written of learning as acquiring knowledge, as knowledge and wisdom as something you "get." The acquisition metaphor is so omnipresent that most people, including us, use it most of the time without thinking about what it might entail. When we say a student *has* a good understanding of fractions, we do not stop to ask ourselves whether the student *owns* this knowledge of fractions in the same way that he owns a DVD or his winter jacket. But in the last 20 years, a number of psychologists have begun describing learning not as acquiring knowledge, but as participating in a "community of practice" (Lave and Wenger 1991; Wenger 1998). The participation metaphor allows us to see that learning something involves the learner in *doing* rather than in *receiving*. If we place these metaphors side-by-side, we can see the consequences of both more clearly. When we describe learning through the language of acquisition, the learner can appear as a passive recipient of another's teaching, as one who receives information without necessarily putting forth much effort, much as baby birds receive worms brought home by their parents. The participation metaphor, by contrast, situates the learner as an actor who engages with the community's common task (Sfard 1998). Wenger (1998) elucidates the participation metaphor with several examples—one being that of an insurance adjusters' office where novices engage with processing claims, seeking others' aid when they encounter an unfamiliar claim, situation, or quandary.

Conceptualizing thinking and learning as participation and communication allows us to see both as social processes. We learn with and from others, through observation, shared work, and conversation. As we explain our thinking to others, we deepen our own understanding. As we try to convince others of the wisdom of our ideas, we may locate gaps in our own logic. Vygotsky (1978) describes the ways in which language actually helps children solve problems and notes that physical and cultural tools often mediate our ability to build understandings. Complex instruction builds on these understandings by providing a structured classroom practice that maximizes children's opportunities to use language and tools to make sense of mathematics.

In *The Power of Their Ideas,* Deborah Meier (2002), the founder of three extraordinary urban public schools, notes that over the course of many decades spent in schools she has discovered that "teaching is mostly listening, and learning is mostly telling." Other educators have been coming to similar conclusions. This lesson has been hard for most adults to absorb, however, because the schools we attended —and many schools even now—taught us that learning was something we did silently, as we listened to our teachers explain how to find a least common denominator, and by ourselves, as we circled the pictures that rhymed with cat, outlined the main ideas in the history book, and solved the even-numbered math problems on page 47. Our teachers saw the social part of school, whether it was recess, lunchtime conversation, or whispers and note passing that happened during lessons, as a distraction from academic learning, which it usually was. Recent work in teaching and learning, however, tells a different story, saying that if we want our students to come to robust understandings of hard ideas, we must let them talk, and we must teach them to listen to and reason about their classmates' ideas.

In the past 25 years, the National Council of Teachers of Mathematics (NCTM) has provided inspiring glimpses of what can happen when teachers make substantial room in math classrooms for communication about mathematical ideas. For example, *Principles and Standards for School Mathematics* (NCTM 2000) offers a snapshot of an elementary school class seeking ways to combine 1.14, .089 and .3 grams of gold. The students, not having received an algorithm for performing this computation, struggle to make sense of the numbers and the task. The teacher poses questions that challenge their assumptions, making a space for developing deeper understandings of decimal numbers than the children have gotten from their earlier school work. Teachers and student teachers that we work with are often powerfully drawn to these images, but they struggle to realize this vision in their own classrooms. Some lament, for example, that repeatedly, when they pose challenging problems, a few, specific students volunteer ideas while their classmates wait for these students to solve the problems. We have, in the past, worked extensively with elementary and middle school teachers struggling with problems of this sort. Fortunately, some excellent resources are available to teachers who are working on skills related to orchestrating good discussions of mathematical ideas (e.g., Chapin, O'Connor, and Anderson 2003; Heaton 2000; Lampert 2001; TERC 2006).

Full-class discussions of math problems should play an important role in math teaching. They give teachers a way to teach children that math is about ideas, to highlight connections and misconceptions and examine them publicly, and to teach norms of intellectual discourse—respectful listening, thoughtful consideration of others' suggestions, and polite, reasoned disagreement. But because having all children continually contribute to a full-class discussion is clearly impossible, many teachers have responded to NCTM's call for change by suggesting that their students work on math problems in groups. When a class of elementary or middle school students work together in groups of two to four, more children have an opportunity to talk at any one moment than would in a full-class discussion. The buzz we hear when we walk into a classroom where children are working in groups makes the truth of this analysis abundantly clear. Moreover, when children are working in groups of three or four, they cannot sit back and wait until Carlos, who *always* has the answers in math, comes to the whiteboard and explains his thinking. They must instead attack the problem using their own intellectual resources and those of their groupmates. In other words, working in small groups or with a partner can create the need for each child to engage with the problem. It can also create a larger space for each child to explain his or her ideas and confusions. When we frame learning as participation, we see that working in small groups on a common problem could offer children unparalleled opportunities for learning, because groupwork opens up more opportunities for active participation.

As we have said before, however, groupwork also creates challenges. Teachers who are trying to use groupwork to deepen all their students' mathematical understandings have limited resources available. This book aims to fill that gap. We also highly recommend Elizabeth Cohen's *Designing Groupwork* (1994). Cohen, however, looked at groupwork as a general pedagogical strategy. We build on her work, but unlike her, we focus on groupwork's place in mathematics teaching and learning.

Many children and adults enjoy working in groups. They appreciate the opportunity to talk while they work, even if not always about the academic task assigned them, and to get help quickly when they are confused. Like LaDonna, a fourth grader (Featherstone 1996), they find working in groups "much *funner*" than working alone.

"I like working with other kids because I finish my work," LaDonna explained to a visiting videographer. "We all work on it together, and you concentrate. And when I work by myself, I don't finish my work" (Featherstone 1996, p. 22).

Groups do not, however, work well for all children, as we saw so clearly in Elise Murray's story. Until Elise made Annette visible to her three classmates, Annette's group excluded her entirely from their work. Observations such as Elise's have led many teachers to abandon groupwork. But our experience indicates that when teachers recognize unequal status as an issue; use pedagogical strategies that address it; design mathematically rich, groupworthy tasks; and build both individual and group accountability into assignments, groupwork offers students powerful opportunities for mathematical learning. The ideas that undergird complex instruction can help teachers make this happen.

The discovery that complex instruction could help teachers enable rich mathematical learning for all students in highly diverse classrooms spoke powerfully to this book's six authors. You have read Lisa Jilk's story in the book's prologue. The rest of the authors—Amy Parks, Marcy Wood, Sandra Crespo, Joy Oslund, and Helen Featherstone—learned about complex instruction from Lisa (more on this later in this introduction). The possibility that complex instruction could help us to do a better job preparing prospective teachers to teach the full range of learners in their classrooms—those who struggled with math, those who were learning already, and those who finished their math work quickly and seemed to need more challenge—excited us immediately.

Beliefs about how you get to be successful in math differ from culture to culture; those in the United States differ markedly from those in Asian countries (Stevenson and Stigler 1992, p. 8). American culture teaches that some people are "math people" and others are not—that you either have what it takes to do math with pleasure and success, or you don't. We do not know why these ideas are so pervasive, but our experiences in grades K–12 and college classrooms convince us that the narrow definition of math that determines what happens during math class in U.S. elementary schools contributes to the American notion that math aptitude is somehow fixed at birth. When being "good at math" is defined, at least de facto, as recalling math "facts" quickly and working out accurate answers to computation problems, only a few children will look like "math people." When teachers and children expand their definition of math to include reasoning, problem solving, and communication, as NCTM (1989, 1991, 2000) has urged in recent decades, "math" calls for more of the intellectual tools that mathematicians use routinely and begins to require a more diverse array of skills than those we see at work in most elementary school math classes. When that happens, children like Annette get a chance to demonstrate what they know.

Let's look, for example, at the spatial-reasoning problem that Elise Murray's fourth graders were struggling with at the beginning of this introduction. Elise asked the children to work with others in their groups to build a block structure that satisfied all the following conditions (adapted from Erickson [1989]):

- There are six blocks in all.
- One of the blocks is yellow.
- The two red blocks do not touch each other.
- The two blue blocks do not touch each other.
- Each blue block shares one edge with each of the red blocks.
- The green block shares one face with each of the other five blocks.
- Each red block shares one edge with the yellow block.

Although we don't know for sure what happened before Elise joined Annette's group, it seems likely that the boys read the clues in sequence, chose a yellow block and then two red blocks from their block bag, then perhaps placed the red blocks on either side of this yellow block, and the blue blocks at each end of the little red and yellow train they had formed. A step-by-step approach of this sort works well for most fourth-grade math assignments. In this problem, however, the approach would have foundered when the boys reached the fifth condition. They could not have satisfied this constraint without disrupting the lineup they had created. This task requires some spatial imagination and a capacity to keep multiple conditions in mind simultaneously. We do not know exactly how and why Annette saw what her three groupmates had missed, but we do know that this task calls forth mathematical strengths that the curriculum in many elementary school classrooms leaves untapped. In doing so, it teaches the fourth graders that math is more than arithmetic. It also repositions Annette as a math student: when a popular, academically successful boy labels her contribution "so smart," he tells his classmates that they could learn from Annette, and he tells Annette that she can succeed in learning math. Because his exclamation clearly derives from observable fact—Annette's contribution *did* enable the group to solve the problem—she has cause to believe him. Further, realizing that she is smart gives Annette reason to work hard in math class.

The spatial-reasoning task may have given Annette an especially good opportunity to display her mathematical talents, but mathematical tasks involving numbers and computation can be just as rich and just as groupworthy. Teachers use complex instruction to teach arithmetic as well as other strands of math. In chapters 4 and 7 and appendix B, we offer the reader guidelines for identifying and creating good tasks for groups. We also discuss ways to modify problems found in ordinary math textbooks, citing some specific examples of tasks we have modified so that they work well for groups.

Many—perhaps most—of the prospective teachers taking our college classes have come to believe that they are not "math people." The experiences that have led them to this conclusion make it easy for them to draw similar conclusions about children in their classrooms who struggle to learn math facts and seem to have little sense of how our number system works. Our rhetoric did not in the past persuade them otherwise; what they needed was to *see and hear* students they thought of as "low" in math making substantive intellectual contributions to math problems. They needed to see success where they did not expect it.

And so the rest of us became excited when we began to learn from Lisa Jilk how she and her colleagues at "Railside High School" in California had used complex instruction to improve learning opportunities for poor and immigrant teenagers in their school. Lisa tells this story in the prologue. (Jo Boaler and her colleagues at Stanford University assigned the school the pseudonym *Railside High School* in their papers describing the research they did there [Boaler and Staples 2008]. The school is, in fact, located next to a railroad track.) Railside's students were very successful in mathematics: 41 percent were taking precalculus or calculus by their senior year, in contrast to 27 percent in the two comparison schools, both of which served somewhat more affluent populations (Boaler and Staples 2008). This success, along with Lisa's enthusiasm and expertise, encouraged two of us to integrate work on complex instruction into math education courses for prospective elementary school teachers at Michigan State University.

Lisa worked with the elementary school student teachers across three three-hour seminar periods to help them design a "complex-instruction math lesson" that they would teach in their own classrooms. The experiment's results exceeded our wildest hopes. The student teachers had left the third seminar nervous and grumpy at the prospect of teaching a math lesson in an entirely new way. However, they returned to class the following week to report excitedly that they had seen children who had never before volunteered an answer in math class participating eagerly in discussions in their small group. One student teacher wrote,

My students simply came alive through this lesson! It was as if I was teaching a whole different class! What was more amazing was the fact that each and every one of my students seemed to stand out. It is difficult for me to describe this in words, but in other words the voice of those students who always seemed to be more or less invisible, whether because they are passive or unmotivated, could suddenly be heard!

The students were grouped on the carpeted floor in different areas of the classroom. Engagement and excitement were radiating off of them through their body language. They were up on their knees, their body was upright facing the center of the group, and their heads were together.

I had purposely assigned the role of Materials Manager to those children who did not have high status in the classroom, especially during group work. (*Authors' note*: Teachers using complex instruction assign each student in a group a role. See chapter 3 for a detailed treatment of what the roles are, what they afford, and how best to introduce and use them). We followed two students in particular, KOD and VIN. The assignment of roles was a big success in pushing KOD higher in status. He was exhibiting a heightened level of confidence through his body language. His chest was puffed up in a dignified position whereas usually he is slumped down sitting outside of the group, doodling. He had the full attention of the group, and he was extremely focused on getting the team's questions just right. His motivation was also improved as, for once, he was not the last one to be done!

The student teachers realized that Lisa had shown them a new way to see children who were doing poorly in math. Describing these children as "having low status" instead of as "my low kids," "low achievers," or even "struggling students" helped the student teachers see these students as ones who could learn mathematics, rather than as ones who would never do well in it. Rethinking their views of these "low" students allowed the student teachers to find more effective ways to open math up to *all* their students. (See Crespo and Featherstone [in press] for more about what it means to change in this way the lens through which prospective teachers look at differences in children's math skills.) Perhaps equally important, several student teachers noted that students who had always done well in math commented on how challenging the task had been and on how hard it had made them think. Complex instruction could give elementary and middle school teachers a way to achieve equitable outcomes in math teaching, just as it had done for the secondary school teachers at Railside High School.

In addition, we began holding complex instruction workshops for the practicing teachers who worked with our students. Excited as we were by the ways in which the teachers and prospective teachers used what they had learned in the complex instruction workshops in their elementary and middle school classrooms, we quickly realized two things. First, we could not, in one week, cover everything that teachers needed to know. Second, and perhaps more important, we could not expect anyone to make sense of all that they did in the workshops so quickly. We felt strongly the need to provide teachers and prospective teachers with some sort of reference that focused specifically on the problems and possibilities that come with a serious effort to use complex instruction in elementary school math teaching.

The result is this book.

In the pages that follow, we lay out what we have learned about using groups to teach math conceptually to all children. We explain how complex instruction opens doors for children now seen as "low" or unsuccessful in math and how it can also support the learning of children who find little challenge in their current math curriculum. We analyze what it takes to design assignments that support children's mathematical learning in groups, what we and others call *groupworthy tasks,* offering readers examples of good tasks and help in adapting math problems they find in their own curricula. Chapter 1 addresses the narrow conception of math smarts that prevails in most schools and communities, asking what it means to be smart in math, explaining how and why we might want to define math smarts differently, and laying out some ways to do this. Chapter 2 examines the issue of status and how it affects students' learning in math class. Chapter 3 describes how teachers lay the groundwork for equitable participation by teaching their students to play carefully defined roles and setting particular norms for groupwork. The nature of the tasks given to groups is as important as the teacher's pedagogical strategies for addressing status issues. Chapter 4 looks closely at one groupworthy task, showing how, when used in the context of a classroom organized for complex instruction, it can call forth a variety of intellectual strengths, introduce important mathematical ideas, and lay the foundation for equitable engagement and learning. Chapters 5 and 6 describe pedagogical strategies for countering the effects of status in groupwork. Chapter 7 examines curricular issues and offers suggestions and examples for teachers who seek to create groupworthy mathematics tasks. Chapter 8 brings in voices of teachers who are using complex instruction as a significant part of their math program. These teachers lay out for readers both the challenges they have encountered and the rewards they have experienced as they tried to put complex instruction's ideas into practice.

We want to conclude this introduction by noting that although our own work has focused on the ways in which complex instruction enables teachers to teach challenging math in highly diverse classrooms, teachers with whom we have worked, and many others across the United States, have applied the principles that undergird complex instruction to other parts of their own curriculum, planning groupworthy tasks for social studies, science, and literature-based inquiry. A sixth-grade teacher in East Lansing, Michigan, reports that her students work so well in groups that she uses them in all parts of her curriculum. These reports hearten us. We encourage readers to consider adapting complex instruction to their own curricula.

References

Ben-Ari, Rachel, and Peri Kedem-Friedrich. "Restructuring Heterogeneous Classes for Cognitive Development: Social Interactive Perspective." *Instructional Science* 28, no. 2 (March 2000): 153–67.

Boaler, Jo, and Megan Staples. "Creating Mathematical Futures through an Equitable Teaching Approach: The Case of Railside School." *Teachers College Record* 110, no. 3 (March 2008): 608–45.

Brown, John Seely, Allan Collins, and Paul Duguid. "Situated Cognition and the Culture of

Learning." *Educational Researcher* 18, no. 1 (January 1989): 32–42.

Chapin, Suzanne H., Catherine O'Connor, and Nancy Canavan Anderson. *Classroom Discussions: Using Math Talk to Help Students Learn, Grades 1–6*. Sausalito, Calif.: Math Solutions Publications, 2003.

Cohen, Elizabeth G. *Designing Groupwork: Strategies for the Heterogeneous Classroom*. 2nd ed. New York: Teachers College Press, 1994.

Cohen, Elizabeth G., Rachel A. Lotan, Percy L. Abram, Beth A. Scarloss, and Susan E. Schultz. "Can Groups Learn?" *Teachers College Record* 104, no. 6 (March 2002): 1045–68.

Crespo, Sandra, and Helen Featherstone. "Counteracting the Language of Math Ability: Preservice Teachers Explore the Role of Status in Elementary Classrooms." In *Mathematics Teacher Education in the Public Interest,* edited by Laura J. Jacobsen, Jean Mistele, and Bharath Sriraman. Charlotte, N.C.: Information Age Publishing, in press.

Erickson, Tim. *Get It Together: Math Problems for Groups, Grades 4–12*. Berkeley, Calif.: EQUALS, 1989.

Featherstone, Helen. "If Everyone Had a Pencil, Chantelle Wouldn't Learn Alex's Strategy." *Changing Minds* 10 (1996): 17–25.

Frey, Nancy, Douglas Fisher, and Sandi Everlove. *Productive Group Work: How to Engage Students, Build Teamwork, and Promote Understanding*. Alexandria, Va.: Association for Supervision and Curriculum Development, 2009.

Heaton, Ruth M. *Teaching Mathematics to the New Standards: Relearning the Dance*. New York: Teachers College Press, 2000.

Janssen, Jeroen, Femke Kirschner, Gijsbert Erkens, Paul A. Kirschner, and Fred Paas. "Making the Black Box of Collaborative Learning Transparent: Combining Process-Oriented and Cognitive Load Approaches." *Educational Psychology Review* 22, no. 2 (June 2010): 139–54.

Lampert, Magdalene. *Teaching Problems and the Problems of Teaching*. New Haven, Conn.: Yale University Press, 2001.

Lave, Jean, and Etienne Wenger. *Situated Learning: Legitimate Peripheral Participation*. Cambridge, U.K.: Cambridge University Press, 1991.

Meier, Deborah. *The Power of Their Ideas: Lessons for America from a Small School in Harlem*. Boston: Beacon Press, 2002.

National Council of Teachers of Mathematics (NCTM). *Curriculum and Evaluation Standards for School Mathematics*. Reston, Va.: NCTM, 1989.

———. *Professional Standards for Teaching Mathematics*. Reston, Va.: NCTM, 1991.

———. *Principles and Standards for School Mathematics*. Reston, Va.: NCTM, 2000.

Piaget, Jean. *The Language and Thought of the Child*. New York: Routledge, 1959.

Sfard, Anna. "On Two Metaphors for Learning and the Dangers of Choosing Just One." *Educational Researcher* 2, no. 2 (March 1998): 4–13.

———. *Thinking as Communicating: Human Development, the Growth of Discourses, and Mathematizing*. Cambridge, U.K.: Cambridge University Press, 2008.

Stevenson, Harold W., and James W. Stigler. *The Learning Gap: Why Our Schools Are Failing and What We Can Learn from Japanese and Chinese Education*. New York: Summit, 1992.

TERC. *Investigations in Number, Data, and Space*. Boston: Pearson, 2006.

Wenger, Etienne. *Communities of Practice: Learning, Meaning, and Identity*. Cambridge, U.K.: Cambridge University Press, 1998.

Vygotsky, Lev S. *Mind in Society: The Development of Higher Psychological Processes*. Cambridge, Mass.: Harvard University Press, 1978.

1

What Does Being Smart in Math Mean?

I F WE WANT children to engage seriously with the task of understanding mathematics and not just strive to learn the procedures for getting correct answers, we must convince them that they *can* learn rigorous math. For even if we, their teachers, carefully avoid using words like *smart* and *dumb* to describe children or their problem-solving strategies, students will often label themselves and their classmates with words like *smart, stupid,* or *lazy*. Sometimes this labeling is silent, but negative labeling, even when unspoken, has consequences. A girl who believes she is smart at math has far more reason to struggle with a difficult idea or a challenging problem than a child who believes she is not smart. The child who believes she is smart in math is more likely to expect to succeed in solving the difficult problem, and she is more likely to believe that she will learn something by struggling with it.

Some educators have protested our use of the terms *smart* and *smartness*. We insist on using these words, because even if adults avoid applying them to youngsters, children want to know that they are smart. Their beliefs about their own smartness, or lack of smartness, have an important impact on their mathematics learning. Moreover, children do not depend on their teachers to tell them whether they are smart: they monitor their own experiences doing math, and they pay attention to how others treat them.

In most elementary school math classrooms, teachers and students alike tend to assume that being smart in math means knowing the procedures for getting accurate answers to computation problems quickly. Many students and teachers include "knowing the facts"—addition, subtraction, multiplication, division—as an important attribute of a smart math student. Although it is certainly true that solving arithmetic problems rapidly and correctly is a useful mathematical skill, two good reasons exist for teachers and students to broaden and enlarge their ideas about what being smart in math means. First, when we define math smarts narrowly, only a few people will believe they are smart in math, and only a small number will look smart to others. In consequence, not many children will enjoy math and put all their intellectual resources to work on challenging math assignments. Second, this narrow vision of smartness as computational speed communicates an incorrect image of what math is and of what it means to "do math." It limits the math curriculum and children's opportunities to learn. These are both good reasons for expanding our ideas about what being smart in math means. We will examine them one at a time, starting with the first: how a narrow vision of math smarts limits the likelihood that most children will come to see themselves as smart in math and engage challenging math problems with confidence and intellectual imagination.

Opening Up New Ways to Be Smart in Math

When math is just computation, only a few children—those who add, subtract, multiply, and divide easily and accurately—will feel smart in math. Opening up more ways to do math creates opportunities for more children to see themselves, and for others to see them, as smart in math. We want to illustrate this point by looking at some of what the prospective teachers in one of our classes wrote when we asked them to identify ways that they were smart in math. Some, of course, maintained that they were *not* smart in math, period. But we insisted that they dig deeper into their experiences with math, looking for those moments or spaces in which they had believed themselves competent in doing math. Here is a random sample of the list one class made:

- I'm good at asking questions to catch up.
- I'm good with guess and check.
- I'm smart at math because I am logical.
- I know my times tables.
- I'm good at math because I'm motivated to solve problems.
- I am open to new approaches.
- I am smart in math because I'm good at multiplication.
- I can balance my checkbook.
- I am comfortable using graphs and tables.
- I am good at making charts.
- I am visual-spatial as a learner and can visualize shapes in 3-D.
- I can follow directions.
- I am smart at estimating.
- I can make predictions.
- I work well in groups.
- I can explain my thinking.
- I can use multiple strategies to solve one problem.
- I can think under pressure.
- I can visualize how problems work before I solve them.

Not surprisingly, some items on this list—16 percent of the full list, from which we randomly selected the items above—suggest success with at least one aspect of computation. That is as it should be: computation is an important part of the elementary school math curriculum, and some children compute rapidly and accurately. Often, this accomplishment makes them feel smart and helps them enjoy math and feel confident in attacking challenging math problems. Most of the prospective teachers did not, however, claim to be good in any aspect of computation. Forty-eight of the 57 items on the prospective teachers' list offer a rich menu of ways that students can feel smart at math even if they make mistakes when multiplying or subtracting. Although some entries, such as "I am good at algebra" and "I am good at trigonometry," presumably derive from high school or college experiences, most are smartnesses these young people might have put to good use in elementary school math lessons. Yet few of these prospective teachers felt successful in elementary school math. Most discovered their math smartnesses later on, if at all. When they looked beyond computation, these prospective teachers found that they were smart in diverse ways.

One goal of complex instruction is to extend children's ideas about what sorts of interests, skills, and dispositions can help to make them smart in math. One of us remembers a relevant experience from the summer she turned ten:

> I grew up in the country, and only a few children my age lived within walking distance. So I passed most of the incredibly hot summer afternoons in playing games of various sorts—softball, croquet, Monopoly—with my brother, Steve. He was four years older than me, and he always won these games. By July, losing games had grown tiresome.
>
> And so I was not altogether resistant, one summer afternoon, when Steve proposed to teach me something. On the one hand, the assumption of superiority irritated me a little; on the other hand, he did seem to know more than I did, and the idea of falling formally into the role of student sounded a little less tiring than losing at Monopoly. However, I was entirely skeptical, because what he wanted to teach me was math. What was there to learn about math? With pretty near complete confidence, I announced that I could solve any math problem he could give me. After all, I knew how to add, subtract, multiply, and divide; I had "had" fractions. I had never seen or heard of a math problem that required anything more than a sharpened pencil and vigilant attention to detail.
>
> Steve came back at me immediately with a problem that permanently changed my assumptions about what math was and what I knew of it. Fifty years later, I remember both the problem and my own astonishment. "John is three times as old as Mary," he began, grinning impishly. "In five years John will be twice as old as Mary. How old are John and Mary?"
>
> He had silenced me. This problem was not like any I had ever encountered. I saw no way to add, subtract, multiply, or divide these numbers to get a sensible answer. Both John's age and Mary's seemed to depend on information that the problem neither provided nor gave a way to figure out. Everything depended on things I hadn't been told. I tried trial and error: what if Mary was 8 and John was 24? No, that wouldn't work, because in 5 years Mary would be 13 and John would not be 26. I tried again. And again. After a while I reluctantly gave up. Steve had won: he had given me a math problem that I couldn't solve.
>
> I was dismayed, but I was also curious. Could there actually be a way to solve a

problem like this one? It didn't really seem to be a math problem: it was more like a riddle. I saw, of course, that Steve could make up a problem like this: he could pick the numbers first and then choose the hints he gave me. But was there really a way to figure the answer out if you weren't a mind reader? Steve claimed that there was, and that he could teach it to me. I think perhaps I was skeptical enough to suggest that I could give *him* such a problem and see whether he could find the numbers I was thinking of. I'm not sure if that happened, but in any case, he did show me a way to attack the problem he had given me. He wrote something along these lines:

$$x = \text{Mary's age now}$$
$$y = \text{John's age now}$$
$$x + 5 = \text{Mary's age in 5 years}$$
$$y + 5 = \text{John's age in 5 years}$$

and

$$y = 3x$$
$$y + 5 = 2(x + 5).$$

Then he showed me that I could combine the two "equations" (this was a new word for me) in such a way that there was only *one thing* I did not know:

$$(3x) + 5 = 2(x + 5)$$

I could then simplify this equation, first by doing the necessary multiplication,

$$3x + 5 = 2x + 10,$$

and then by subtracting $2x$ from both sides of the equation. I was left with

$$x + 5 = 10.$$

My experience with arithmetic made the next step easier:

$$x = 5.$$

Now, he explained, I knew Mary's age. How old is John? How old will they both be in 5 years?

I doubt that this conversation was accomplished in a single afternoon, but I do remember that I was astonished when I saw how such a problem could be solved and that I enjoyed the detective work involved in finding these answers.

We played this game off and on for a week or two, alternating with croquet, softball, and Monopoly. I rather enjoyed it. It was less exhausting than playing croquet or softball in the broiling sun, and less frustrating than watching my money trickle away in endless Monopoly games. It also provided me with a magic tool for puzzling my friends.

Steve told me that this was a kind of mathematics called *algebra,* and that I would study it in high school. I thought it a vast improvement on the arithmetic we were doing in school, since there were no tedious computations and fewer opportunities to make what my teachers called "careless" mistakes. And I loved the fact that when I had an answer

I could plug the numbers back into the original story and show myself that I was right. Besides, it had an element of magic: the possibility of discovering two different numbers, when all we knew about either was given in terms of another that we did not know, seemed an incredible sleight of hand.

Here we see a girl who has never felt smart in math glimpsing for the first time a future in which she will engage math with success and pleasure. Doing a little algebra outside school changed her ideas about what math was and what it took to be good at it. In the context of problems like this one, willingness to experiment, to think in unfamiliar ways, and to make and revise educated guesses became useful math smartnesses. Expanding the meaning of "doing math," even for a few summer afternoons, allowed this ten-year-old to see that she did, after all, have some math smartnesses. She went on, in fact, to do well in high school math and Advanced Placement calculus.

Experiences like this one probably made us particularly receptive to the revisioning of school mathematics's goals proposed by the National Council of Teachers of Mathematics (NCTM 1989, 2000). This revisioning proposed goals based on what math educators know about the nature of work in the mathematics discipline. According to NCTM (2000), math instruction in grades K–12 should enable children to do the following:

- Make and investigate mathematical conjectures;
- Develop mathematical arguments;
- Make sense of the mathematical arguments others make;
- Evaluate [the logic in] mathematical arguments made by others;
- Develop new mathematical knowledge through problem solving;
- Solve mathematics problems that occur both inside and outside math classes;
- Find and create strategies for solving a new problem;
- Describe and reflect on strategies they have used for solving a problem;
- Communicate their ideas about solving a math problem to others;
- Extend and consolidate their mathematical knowedge through communication with others;
- Use the language of mathematics to communicate ideas;
- Make connections among mathematical ideas, and use these connections to solve problems and generate new ideas;
- Connect and apply mathematics in contexts outside mathematics;
- Create and use representations in order to organize, record, and communicate mathematical ideas;
- Translate among mathematical representations;
- Select and apply representations in order to solve problems; and
- Use representations to model physical, social, and mathematical phenomena.

This list clearly opens up many ways for a student to be smart in math. It also reminds us of two important points about children's math smartnesses. First, math smarts are not abilities a person is born with: one builds them through use. Like expertise at chess or piano playing, math smarts develop over time as children engage with challenging problems that call them forth. Since different people lean toward different ways of thinking and solving problems, children develop particular math smarts differently and on different timetables. If we look at the skills on NCTM's list above, we see that although some people will find it easier than others to develop particular skills, nearly all these skills will develop with use. As teachers, we want to help students to appreciate the smarts they already have while we support them in developing new ones. This leads us to a second point: *Smart* is not an all-or-nothing attribute. We are all smart at some things we have practiced over the years; we will all get better at these and other things as we spend more time trying to do them. We want to enlarge children's mathematical opportunities beyond computation because, as we saw in the story about a ten-year-old's encounter with algebra, if we provide these expanded mathematical opportunities, a child who has long believed herself "unsmart" doing three-digit computations may learn that she can potentially develop significant smarts in some other sort of math.

But if we take NCTM's list seriously, we need a very different kind of mathematics program than the one most adults encountered in elementary and middle school. Doing "mad minutes" (i.e., timed tests of math facts, like $3 + 5 = 8$ and $7 \times 6 = 42$) and working on a sheet of two-digit addition or division problems may seem like a good way to motivate children to memorize and develop speedy recall of math facts. Research (e.g., Baroody [1985]; Rathmell, Leutzinger, and Gabrielle [2000]; Thornton [1990]), however, indicates that these methods do not actually help much (teaching children strategies for figuring out the math facts they can't remember seems to be more effective) and will affect most of the skills on NCTM's list only very indirectly. Even working alone on the word problems at the end of a textbook chapter has limited value in helping children to develop the skills NCTM has identified as valuable. Because these problems usually require applying computational techniques taught earlier in the chapter, children need only turn to a previous page in their textbook to find a winning strategy for solving them. Often, these problems simply circumscribe in the student's mind the situations in which the skill just taught will be useful.

Creating Opportunities for Children to Be Smart in New Ways

NCTM's list of skills may seem a bit abstract. What might working seriously on them look like in practice? To begin to think about this, let's take the hypothetical case of a group of seventh graders who have just learned how to divide a mixed number—a number like $1\frac{1}{4}$ that is made up of a whole number and a fraction—by a fraction, and whose teacher has given them the assignment in figure 1.1.

This task is far harder than it looks. The difficulty resides in the fact that "divide *by* one-half" sounds very much like dividing *in* half, something that we do almost every day but is actually quite different. Half an hour ago, Helen divided the stale brownie on the kitchen table in half so that she and a friend would each have something sweet to eat while they sipped their coffee. Dividing the brownie in half is the same as dividing it *by 2*: it results, if done skillfully, in two equal pieces. Similarly, dividing *by 3* results in three equal pieces, and less stale brownie for everyone. So, following this logic, what could dividing *by one half* possibly mean? Does that even make sense? Well, if you stick with the image of cutting up brownies, it really doesn't.

Dividing by a Fraction

Our textbook says that to divide a fraction by another fraction you "change mixed numbers to improper fractions, and then invert the denominator and multiply." So to do the problem $1^3/_4 \div 1/_2 = ?$, we change $1^3/_4$ to $7/_4$, flip the $1/_2$ over to get $2/_1$, and multiply $7/_4$ by $2/_1$. The answer is $14/_4$, or $3^1/_2$. But even though you divided, $3^1/_2$ is bigger than the number you started with. Explain how this could make sense in a way that a fifth grader could understand. Give examples of situations in which it would make sense to divide by a fraction. Use pictures, words, and symbols.

Fig. 1.1

Making sense of a request to divide by one half or any other fraction is possible, however, if we think a bit more about the various ways we can use division, excluding fractions from the conversation at first. Most of us—children and adults—think about division as meaning separating an object or collection of objects into a given number of equal parts (i.e., We have a bag of 20 M&M's to divide among 4 people: how many will each of us get?). This is, indeed, an important meaning of division. We call it the *partitive* meaning, because it involves dividing something into equal parts. But it is not the only meaning of division. We also commonly use division to solve problems like "It takes 2 eggs to make french toast for one person; how may people can have french toast this morning if I have 12 eggs?" When we divide 12 by 2 to answer this question, we are invoking the *measurement* meaning of division. We are measuring our egg collection out into six two-egg servings. So when we say "12 ÷ 2 = 6," we can either be thinking of creating two sets of six eggs—the *partitive* meaning (fig. 1.2)—or six sets of two eggs each—the *measurement* meaning (fig. 1.3).

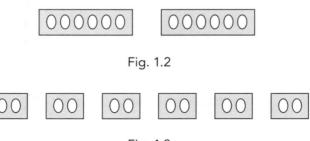

Fig. 1.2

Fig. 1.3

None of this is particularly troublesome when we use small whole numbers, as we do here. Most of us use the measurement meaning of division from time to time. For example, yesterday one of us counted the small candies she had bought for Halloween trick-or-treaters and found that she had 61. She used division to answer this question: If she wanted to give each child three candies, how many children was she prepared for? However, when we think about division, most of us think first of the more familiar, partitive meaning. Knowing what dividing "by $1/_2$" means is difficult: what would saying "let's share this brownie among half a person" mean? Most children and adults assume that dividing *by* $1/_2$ means the same thing as dividing *in* half. In order to make actual sense of what dividing "by $1/_2$" means, we need to think using the measurement meaning of division. So we might illustrate "$1^3/_4 \div 1/_2$" with a story like the following:

- ❑ Your parents are going out for dinner, and your mom says that you can invite friends over to eat the pizza that is left from last night. You decide that $1/_2$ a pizza would be enough for one of your friends. It turns out that $1^3/_4$ pizzas are in the refrigerator. How many of your friends can you serve?

- ❑ *Answer:* 3 people—you plus 2 friends—with half a serving, or $1/_4$ of a pizza, left over

A drawing for the problem shows $1^3/_4$ square pizzas divided into $3^1/_2$ half-pizza portions, one half-pizza portion for each diner. Figure 1.4 shows each child's half-pizza portion in a different shade.

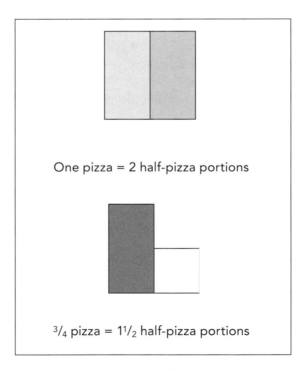

One pizza = 2 half-pizza portions

$3/_4$ pizza = $1^1/_2$ half-pizza portions

Fig. 1.4. A representation of $1^3/_4 \div 1/_2$: $1^3/_4$ pizzas
divided into $1/_2$-pizza servings

Another problem that one of our students created also illustrates very nicely what dividing by one-half means:

□ You are at the amusement park, and your mother tells you that it will be time to leave in $1^3/_4$ hours. You want to spend all the time left riding on the ferris wheel. A ferris wheel ride lasts half an hour. How many rides can you take?

Hardly any of us learned to think this way about division, so most people produce some variant of the following:

□ You and your brother come home from school and find $1^3/_4$ pizzas in the refrigerator. You want to share them equally. How many pieces of pizza will each of you get?

One might explain, "Each of you gets $3^1/_2$ pieces of pizza: you get the dark grey pieces, and your brother gets the light grey ones." Figure 1.5 shows two pizzas divided into four pieces each, with the second pizza missing one piece.

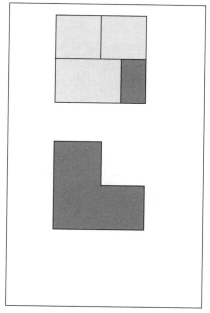

Fig. 1.5. A representation of $1^3/_4$ divided by 2

We would have done pretty much the same ourselves, had we not thought long and hard about this problem. Nearly everyone remembers to "invert and multiply" when dividing by a fraction, as the middle school couplet "yours not to reason why, just invert and multiply" prompts, but few adults can reason their way to an explanation for the rule.

Let's go back to the task we designed for our hypothetical seventh graders: explaining to a slightly younger child what dividing by a fraction means. Most people would agree that any middle school students who have learned how to perform the computation $1^3/_4 \div 1/_2$ ought to be able to do what that fraction problem asks. They ought to be able both to make sense of what performing this computation might mean and to give an example of a time when one might do this. Our experience leads us to be pretty confident that few could. But let's think about the skills that the group doing this work would need in order to succeed at this task. We have put in italic type below those that map onto NCTM's list above.

We ask the children not only to craft an explanation of what is going on when you divide a mixed number by a fraction, but also to use words, pictures, and symbols in that explanation. This last request requires them to *select and apply representations in order to solve problems* and to *translate between representations*—no easy task in this instance, as we have just seen. Since they are producing an explanation, they must *create and use representations in order to organize, record, and communicate mathematical ideas.* To the extent that they succeed in doing what we asked, they will need to *use the language of mathematics to communicate ideas.*

In order to make sense of the math, to figure out what dividing $1\frac{3}{4}$ by $\frac{1}{2}$ could mean, the children have to examine what division actually means, as we have just done. If they manage to *make connections among mathematical ideas* in this way, they will have used these connections to solve problems and *generate new ideas.* Using what they know or can figure out about division, they will have forged a new understanding of division by a fraction and deepened their understanding of the division operation.

Like most of the rest of us, a group of four middle school students faced with this sense-making challenge would, very likely, start down several dead-end roads. If, however, some of the group members' skills included *communicating their ideas about solving a math problem to others, making sense of the mathematical arguments others make,* and *evaluating [the logic in] mathematical arguments made by others,* the group's discussions might after a while lead in productive directions.

We have explored this one example at considerable length because we think it helps make vivid how fundamental the skills on NCTM's list are to the intellectual work of doing and learning math.

Now let's look at a problem for younger students working in groups of four (fig. 1.6).

Number Sentences

(Adapted from Deborah Ball's [1992] hypermedia materials)

Each of you should work alone for the first 10 minutes, writing as many number sentences equal to 10 as you can (for example: 6 + 4 = 10, 1 + 9 = 10).

After 10 minutes I will flash the lights, and you should stop writing new number sentences and look at *all* the number sentences that everyone in your group has written. Look for patterns. Write down as many patterns as your group can find. Make sure that everyone in the group understands and agrees to each pattern that you write down and that each person in the group can explain each pattern. Each of you should explain your thinking in your math journal.

Number Sentences—*Continued*

Make a poster that has at least four number sentences from each person in your group. Use these number sentences as well as words, symbols, and pictures to describe the patterns your group found.

If you have time after you have made your poster, talk about this question in your group: *What is the largest number of sentences equal to 10 that anyone could write?* Be ready to explain to the rest of the class both your answer to this question and the answers of others in your group.

Fig. 1.6

Return now to NCTM's list of skills and smarts, and as you read what Duane, Sheena, Maria, and Haziem did with this task, see which of these smarts you see these third graders using.

Here are some of the patterns that Duane, Sheena, Maria, and Haziem wrote down.

- If you write a number sentence that is equal to 10, such as $6 + 4 = 10$, you can make another number sentence by making the first number smaller and the second number bigger.

- If you make the first number one bigger, then you have to make the second number one smaller.

- If the first number is bigger than 10, the number sentence has to be take-away.

- You can start with any number and make a number sentence equal to 10 ($100 - 90 = 10$, $30 - 20 = 10$).

- If you have a number sentence like Duane's ($21 - 11 = 10$), you can make another number sentence by making the first number bigger and making the second number bigger.

The group agreed quickly on the first pattern: they already knew a lot of ways to "make 10" from earlier work, and most of the children had started by writing $9 + 1 = 10$, $8 + 2 = 10$, and so on. No one thought that they needed to make a picture to explain this; they had talked about it in class. However, the children argued a lot about the fourth assertion, "You can start with any number and make a number sentence equal to 10 ($100 - 90 = 10$, $30 - 20 = 10$)." Duane claimed that you could not know whether if you started with "a million, trillion, billion, or "some *huge* number mathematicians do not even know" you "could subtract enough" to get to 10. Sheena said of course you could, "because you can just keep taking things away until you have just 10 left." Haziem and Maria were unsure. Maria said that she could not even think about "really, really big" numbers." Haziem tried to make a picture of what Sheena was saying and got confused.

But then Duane said, "Look, suppose I had a really, really huge pile of pennies with so many pennies that we couldn't even count them in one day, maybe even in one week. Even with a *huge* pile like that, we could still just take pennies out of the pile until there were only 10 pennies left." Haziem said that made sense. After Sheena tried to make a picture, they all agreed that this pattern worked.

Before you read any further, return to NCTM's list of skills and smarts and see which ones you see Duane, Sheena, Maria, and Haziem using. Also, include in your list any significant "math smarts" you see them using that are not on NCTM's list.

Multiple Smarts in the Classroom

It seems clear to us, and research supports our conclusion, that broadening teachers', students', and parents' ideas about math smartnesses enables many more children to recognize that they are, in fact, smart in math. As Jo Boaler (2006) writes after completing the five-year study of 700 students in math classes in three high schools that we mentioned in the introduction, "The narrowness by which success is judged means that some students rise to the top of classes, gaining good grades and teacher praise, whilst others sink to the bottom. In addition, most students know where they are in the hierarchy created." Teachers in two of the high schools in Boaler's study taught math quite traditionally, but in Railside High School (see the prologue), the math department had abolished tracking and created their own curriculum. Working with Elizabeth Cohen and Rachel Lotan of Stanford University, they designed lessons built according to the principles of complex instruction. Those interested in scaling up should note that no one—neither the school nor the district—mandated the reforms that put complex instruction in place in almost all Railside math classes. Instead, the math teachers and various department chairs, who were all concerned about equity and students' understanding of important mathematical ideas, took the initiative. The department chairs and teachers fought hard to get the school's and district's support for the changes they made.

In all classes, from freshman algebra, which all students took, to Advanced Placement (AP) calculus, students worked in small groups on problems that, like "Divide by a Fraction" and "Number Sentences," called for a variety of intellectual skills. Despite multiple socioeconomic, cultural, and linguistic challenges, students at Railside did spectacularly well in learning math: 41 percent of Railside's seniors were taking precalculus or AP calculus at the time the study ended, compared to 27 percent of the two traditional schools' seniors. Given that the schools that taught math traditionally served more affluent communities than Railside, and that ninth graders entering those two schools scored better on math achievement tests than those entering Railside, this achievement is particularly impressive. Also, despite significant differences among the achievement scores of black, white, Latino, and Asian ninth graders entering Railside, by the end of tenth grade, the differences among white, black, and Latino students' scores had disappeared (Boaler and Staples 2008).

When researchers asked Railside students what it took to be successful in mathematics class, some of the answers they heard were the following.

- Justifying work
- Asking good questions
- Rephrasing problems
- Explaining ideas
- Being logical
- Helping others
- Using different representations
- Bringing a different perspective to a problem

Although a group of primary school children would be unlikely to use these words, all these skills are important to solving math problems in *any* group, and all connect to the ways in which NCTM and other reformers have urged us to think about what math is. In interviews with Lisa Jilk (2007), young women who had graduated from Railside asserted that when they entered high school, they had planned to take only the two math classes that they would need in order to graduate. In their freshman and sophomore math classes, however, they felt smart and successful, so they continued to take math for all four high school years. These young women completed AP calculus and went on to college, where they some took more math courses.

The Railside teachers' experience with complex instruction is the foundation for our work with teachers and prospective teachers, first at Michigan State University, and more recently, at the Universities of Arizona, Georgia, and Washington; Alma College in Michigan; and Brandeis University in Massachusetts.

Acknowledging and Highlighting Multiple Smarts

If, in order to do well, children needed *all* the skills we have listed, almost no one would succeed or feel smart in math class. And, indeed, the problems we have offered in the previous section that require a broad mix of skills are harder than those we typically assign to students in third or sixth grade. If we present them, without support or explanation, to children who have, up until that moment, been moving systematically through their traditional textbook, these problems might well cause panic and unproductive frustration. But when we give such problems in a curriculum built on complex instruction to children who have learned to expect to struggle together with some confusion and to depend on one another's ideas, these math tasks create a context for educative conversations and serious learning.

In the Railside classes, and in elementary school classrooms using complex instruction, teachers often launch a problem-based lesson by listing skills that each group will need to complete the assignment. (See the section on multiple-ability treatments in chapter 5 for more on this pedagogical strategy.) They then assure their students that no individual will have all these skills, but that everyone will have some of them. They further assure them that, because different students have different smarts, they will find that they have the intellectual resources they need in their groups to solve the problem, if they make sure to get everyone's ideas at each stage of the work (see Cohen [1994], pp. 122–30). Of course, this is convincing only if it is true. If problems actually require only carefully applied, previously learned rules, and anyone who has mastered arithmetic operations can solve them readily, children will quickly learn to hear the call to solicit and respect everyone's contributions as just one more adult piety. But if problems call for a wide variety of competences like those listed above, groups will not be able to depend solely on Sean, who always gets 100 percent on the Mad Minutes tests. They will need ideas from Cherise, who asks in a whisper whether dividing a pizza "by a half" is really the same as dividing it "in half," and from Francisco, who proposes that the group can double the number of number sentences equal to 10 by turning the addition sentences around (e.g., if $2 + 8 = 10$, then $8 + 2 = 10$). The teacher who overhears these excellent suggestions and calls them to the group's attention helps all the children value the ideas of classmates they had previously ignored. This teacher also ups the mathematical ante, showing the children that solving a math problem often requires going beyond memorized procedures.

Tasting the Pleasures of Doing Real Math

So far we have argued that, by conceiving the goals of math instruction more broadly, as reflected in the NCTM *Standards* documents (1989, 1991, 2000), and by designing tasks and curricula to address these goals, we can give elementary and middle school students opportunities to discover and develop math smarts that they do not know they have if they know math only as computation. Now we turn briefly to our other reason for reconceptualizing what it means to be smart in math: we want to offer children opportunities to develop mathematical ways of thinking and to taste the pleasures that come with doing mathematics in ways that parallel the work of adult mathematicians.

So, what *is* mathematics, and what do mathematicians do in order to discover new mathematics? Keith Devlin (1994, pp. 1–3) provides a historical answer to the question of what mathematics is, tracing its evolution from *the science of number* in 500 BCE; through *the science of number and shape* in ancient Greece; *the science of number, shape, motion, change, and space* after the seventeenth-century contributions of Liebnitz and Newton; to today's *science of pattern*. Apparently Western mathematicians left behind the idea that math is just computation quite a long time ago.

When we begin to see math as patterns, we are on our way to seeing math as the natural intellectual domain of childhood. Children learn language, perhaps the most important intellectual accomplishment of life's first decade, by identifying, almost entirely independently, the patterns that make up their native language's grammar. The mistakes they make tell us that they learn the rules of speech through pattern identification rather than simple imitation, although imitation plays a role in vocabulary development. Think of "I goed …", something most English-speaking children say at one time or another. The preschooler who makes this assertion has probably never heard an adult say *goed*. He or she has likely heard "I went …" hundreds of times, but has learned from analyzing the patterns in what adults say that one adds the -*ed* sound to the end of a word (e.g., we hop*ed*, I lik*ed*, I want*ed*) if one is talking about something that has already happened. Most children enjoy looking for patterns, and when math becomes pattern hunting, they are likely to be pleased. A six-year-old can access some patterns, like the commutative property of addition, which states that when one adds two or more numbers, order does not matter (i.e., $n + m = m + n$).

Imre Lakatos (1976), a philosopher of mathematics and science, contended that mathematical knowledge grows as mathematicians offer their conjectures about what is true and other mathematicians suggest counterexamples that falsify or qualify–or do not—the first conjectures. Many mathematicians disagree with Lakatos, seeing mathematics as a game with agreed-on rules or as discoverable truths embedded in the material world (Hersh 1997). However, most would agree, we think, that although elementary school math programs ought to develop children's facility with computation, real mathematics involves conjectural leaps and experimentation quite different from the work of completing a page of computations.

NCTM's skills list above represents many math educators' ideas about how what adult mathematicians do might map onto the school mathematics curriculum. But although redefining math smarts is essential to broadening opportunities for students to learn math conceptually, it will not, by itself, ensure that all students will learn math with conceptual understanding. In order to make this happen, we must find ways of teaching math that encourage all students to participate in conversations about math problems that engage these smarts, and that get them thinking about ideas central to mathematics. *Participation* turns out to be a crucial word here: students do not learn unless they contribute actively, and as almost all teachers know, getting all students involved in math conversations is no easy matter. In chapters 2, 3, 5, and 6, we address this issue and describe how complex instruction can help teachers get all students involved and thinking about math problems. In chapters 4 and 7 and appendix B, we consider the design of good math tasks, tasks that can potentially involve students with important mathematical ideas while also offering points of entry for students whose skills and ways of approaching mathematics problems differ widely.

References

Baroody, Arthur J. "Mastery of Basic Number Combinations: Internalization of Relationships or Facts?" *Journal for Research in Mathematics Education* 16 (1985): 83–98.

Ball, Deborah Loewenberg. "Video from a Year of Teaching Third Grade." Part of the Mathematics Teaching and Learning to Teach Project at Michigan State University and the University of Michigan, 1992.

Boaler, Jo. "Urban Success: A Multidimensional Mathematics Approach with Equitable Outcomes." *Phi Delta Kappan* 87, no. 5 (January 2006): 364–69.

Boaler, Jo, and Megan Staples. "Creating Mathematical Futures through an Equitable Teaching Approach: The Case of Railside School." *Teachers College Record* 110, no. 3 (March 2008): 608–45.

Cohen, Elizabeth G. *Designing Groupwork: Strategies for the Heterogeneous Classroom.* 2nd ed. New York: Teachers College Press, 1994.

Devlin, Keith. *Mathematics: The Science of Patterns—the Search for Order in Life, Mind, and the Universe.* New York: Henry Holt, 1994.

Hersh, Reuben. *What Is Mathematics, Really?* New York: Oxford University Press, 1997.

Jilk, Lisa M. "Translated Mathematics: Immigrant Women's Use of Salient Identities as Cultural Tools for Interpretation and Learning." Unpublished doctoral dissertation, Michigan State University, 2007.

Lakatos, Imre, John Worrall, and Elie Zahar, eds. *Proofs and Refutations: The Logic of Mathematical Discovery.* Cambridge, U.K.: Cambridge University Press, 1976.

National Council of Teachers of Mathematics (NCTM). *Curriculum and Evaluation Standards for School Mathematics.* Reston, Va.: NCTM, 1989.

———. *Professional Standards for Teaching Mathematics.* Reston, Va.: NCTM, 1991.

———. *Principles and Standards for School Mathematics.* Reston, Va.: NCTM, 2000.

Rathmell, Edward C., Larry P. Leutzinger, and Anthony J. Gabriele. *Thinking with Numbers.* Cedar Falls, Iowa: Thinking with Numbers, 2000.

Thornton, Carol A. "Strategies for the Basic Facts." In *Mathematics for the Young Child,* edited by Joseph N. Payne, pp. 133–51. Reston, Va.: National Council of Teachers of Mathematics, 1990.

Why Isn't Miguel Learning Math?
Status at Work

A TEACHER OF a bilingual third-grade class writes (Shulman, Lotan, and Whitcomb 1998, p. 69):

> Miguel was a shy and withdrawn child who spoke no English and stuttered when he spoke Spanish. His Spanish reading and writing skills were very low, and although math was his strength, nobody seemed to notice. Recently arrived from a small community in Mexico, Miguel lived with relatives—more than 10 adults and three children in a two-bedroom apartment. He came to school hungry and tired, wearing dirty clothes. Shunned by his classmates, who said he had "cooties," Miguel was left out of group activities. Even when he had a specific role [in a group], other members of the group would take over and tell him what to do. Miguel was obviously a low-status student.
>
> When I observed Miguel's group, I saw that the other members simply wouldn't give him a chance. Cooperative learning was not helping him at all. Miguel grew more isolated by the day. Students increasingly teased him, and he was getting into fights and becoming a behavior problem.
>
> One day in May, we were working in cooperative groups building different structures with straws, pins, clay, and wires. I was observing Miguel's group and saw him quietly pick up some straws and pins and start building a structure following the diagram on one of the activity cards. The other members of the group were trying to figure out how to begin their structure and, as usual, were not paying much attention to Miguel. I observed that Miguel had put double straws to make the base more sturdy. He knew exactly what to do, because he had looked at the diagram on the card. In other words, Miguel knew that the task was to build as sturdy a structure as possible, and he understood the principle of making the base stronger by using double straws.
>

Almost all of us have taught some students like Miguel, children whom their classmates shun on the playground and ignore inside the classroom. When we let the children form their own groups informally, no one asks Miguel to join them; those he approaches turn away slightly, signaling to him, even if we adults do not notice, that he is not welcome. When we set up groups more formally, those assigned to a group with Miguel may complain, as Miguel's teacher notes here, that he has "cooties." Or they may simply ignore him and his ideas. This is apt to happen even if our Miguel actually has relevant skills and knows about the subject under study.

Elizabeth Cohen and Rachel Lotan (1997) explain that a process that sociologists call *status generalization* helps us understand why a child like Miguel gets ignored even in settings where he could contribute quite a bit. This term calls our attention to the way in which people take note of characteristics on which people differ (speech, friendliness, clothes, ethnicity, and reading skills, to name but a few), pool their assessments of various characteristics to decide a person's status, and on this basis predict how successful the person will be on a given task. This often happens even if most of the "status characteristics" have nothing to do with the task's requirements or the "smarts" of those involved. Thus, the third graders in Miguel's class did not expect him to have useful ideas about building sturdy structures out of straws because he wore dirty clothes, spoke broken English, and misbehaved. Even when his actions demonstrated an excellent approach to the assignment, the three other eight-year-olds in his group took no notice: they had written off Miguel months earlier.

Cohen (1994, p. 27) defines status as "an agreed-upon social ranking where everyone feels it is better to have high rank than a low rank." Status is local in the sense that a person's status in a particular environment depends on what people in that environment value: a person may have high status in one group—in their church, for example—and low status in another. Many children enjoy summer camp because they have higher status there than they do in school; some who feel invisible in school become stars in camp. Miguel's classmates had assigned him a low status, so although his ideas about building sturdy structures were better than theirs, none of these other third graders thought to ask his opinion or even to look to see what he was doing.

Status is a word most of us associate more with adolescents and adults than with seven-year-olds. When we have talked to prospective teachers about status as an element in the dynamics of elementary school groups, some have been openly skeptical.

"There are no status differences in our school," one student teacher proclaimed. Several others nodded their agreement. But as they began to examine the interactions among children in their classrooms more carefully, most of these prospective teachers changed their minds.

After introducing our student teachers to the notion of status, we send them off to examine its workings in their classrooms. We suggest they begin by looking for an interaction in which they think status plays a role and write an account of what they observe. Many tell us later that they doubted that they would be able to do this assignment: they did not think that status played any role in their classrooms. However, almost everyone returns a week later with a story.

"When I started looking for status," one reported, "I saw it all over the place." Hands went up around the room as other student teachers volunteered concurring tales.

More often than not, status differences in the classroom reflect those in the wider society. Children who are outgoing and socially skillful rank higher than those who are shy and awkward. Boys may rank higher than girls. European American students may rank higher than African American students. Classmates will usually assign a low status to a child like Miguel, who acts out and can't read English. But one cannot trace all status differences to attitudes embedded in American culture: some—Cohen (1997) calls them *local status characteristics*—derive from the school's or classroom's culture, or both. Children watch their teachers to find out what the teachers value and, at least in the primary grades, the children's interpretations of what they see have a bearing on who has high and low status in the schoolroom. The status of children like Miguel, who misbehave frequently, is likely to suffer: seeing their teacher reprimand these children frequently, classmates begin to treat the "Miguels" as though they aren't smart. In fact, however, children who regularly defy the teacher's directives are just as likely as their well-behaved neighbors to understand how to represent data on a pie chart.

As we watched a video of groupwork in which three students bend over the assignment and a fourth leans back, apparently lost in space, we realized that if we had observed this scene in our own classrooms we would have chided the dreamer, urging him or her to "get in there" and share in the work. After learning about Cohen's ideas about status, we now look at such situations more closely. Often, we see that the students who are working on the assignment are subtly—and sometimes not so subtly—excluding those we saw as dreamers. They have decided whose ideas will be useful and have distributed the work accordingly.

Adults often imagine that groups provide a place where the children to whom math comes easily will help those who struggle more. Conversation with a colleague prompted an enormously talented, conscientious teacher who Helen Featherstone worked with for years to investigate how safe her students felt during math discussions. She distributed file cards to the fifth graders and asked them to write about how they would feel about volunteering an idea or answer that they were unsure of during a math discussion. Several hands went up.

"Do you mean in whole-class discussions or in our groups?" the students asked. The teacher had been thinking about full-class discussions, but, curious about the question, she inquired whether it mattered. When many voices insisted that it mattered very much, she was, she reported, a little surprised.

"But I thought, 'Oh, well, I guess that makes sense. It would be scarier to be wrong in front of the whole class than in their cozy little groups where they know each other so well.'" When she reviewed what the children had written on the cards later, she was astonished to find that nearly all felt comfortable about voicing a partially formulated idea or a perhaps-incorrect guess in the full-class discussion, but quite a few felt less willing to take such risks in their small groups. They did not explain why: we are guessing that the children knew that their teacher would make sure that their ideas were treated with respect during the full-class discussion. Apparently, though, some students found ways to demean others in the small groups and to do so subtly that their actions escaped their teacher's vigilant eye.

We recommend that you stop reading for a minute now and think about the students in your classroom. Which ones might have status issues? How does this affect their participation? The students and situations that spring into your mind may be far less obvious than those we saw in the case of Miguel. Try reframing the question: "When my students are working in groups or pairs, which ones seem to hang back, acting as though they are shy or lazy?" If you keep needing to say, "Come on, get to work," to particular students, this may be a clue that they are having status problems, that neither they nor their classmates expect them to contribute much to math work, or even to school work more generally. Behaviors that look like problems in the student are quite often problems of status rankings in the group.

Academic and social status affect participation in classrooms. Their impact is especially marked in cooperative groups where no teacher is available to protect quieter, less visible students from ridicule and to solicit their ideas. When Marcy Wood (2008) analyzed interactions among fourth graders doing math in small groups, she found that students' participation often changed significantly when the teacher joined the group. When elementary and middle school children work in small groups, low-status students participate far less actively than high-status students, even when the source of high status is irrelevant to the assignment (Rosenholtz 1985; Tammivaara 1982).

This link between status and participation is important because considerable research evidence exists linking participation to learning: the more students talk about the ideas under study, the more they learn.

> One of the most robust findings of the research on Complex Instruction is the positive relationship between student interaction in small groups and average learning gains. This finding holds at the classroom as well as the individual level. At the classroom level, the proportion of students talking and working together is a positive predictor of average learning gains; at the individual level, the student's rate of participation in the small group is a significant predictor of his or her posttest scores, holding constant the pretest scores.
>
> (Lotan 1997, p. 20)

That children who talk less and share fewer ideas in the group learn less should be no surprise: most of us deepen our thinking and learn its limitations by trying out ideas. The research on the relationship between interaction and learning that Cohen (1994, p. 8) summarizes is especially relevant to groupwork that tries to engage children with complex, conceptually rich problems, the sort that are most likely to deepen their understandings of important mathematical concepts and broaden their ideas about what it means to do math. In situations in which participants—children or adults—require "resources (information, knowledge, heuristic problem-solving strategies, materials, and skills) that no single individual possesses," those who participate learn most, whereas those who remain silent learn least.

When we as teachers put students in groups to work on math problems, we may hope that the students who know the most will help those who know less make sense of the math and learn new ways to approach it. In fact, however, when we put students into groups to work on an intellectually challenging math task, we are likely to create a situation in which the students whose status is already high do most of the work, and most of the learning, whereas those whose status is low participate and learn much less. Sadly, at least in elementary school classrooms, the socially successful children are those who their classmates already perceive as academically successful.

We must emphasize here, however, that status is not static: children's status often changes when their environment changes. We pointed out earlier in this chapter that many children who are lucky enough to go to summer camp find themselves viewed differently there than they are in school: in most camps, stamina on a hike, expertise in canoeing, courage in the face of uninvited raccoons, and familiarity with camp traditions matter much more than reading skill. The child who is quiet and withdrawn in her reading group may be the leader of the pack at camp. The good news for students and teachers is that quite small changes in the environment can initiate shifts in students' status. In some instances, students have moved from low status to high status merely because the teacher has identified their mathematical skills publicly. However, this potential for shifting status can be quite fragile. We can undermine it unintentionally, if we begin to think about students as actually *being* "low status" or "high status" rather than as *being in situations* where others act as though some students have much to contribute whereas others have little. In order to emphasize that status is dynamic, we will try to talk about students as *having* low or high status rather than as *being* low-status or high-status students. This small shift in language can help us move away from the tendency to see low or high, whether status or ability, as a student's more permanent characteristic.

In a school, curriculum can affect status. A third-grade girl who her classmates and her teacher admired for her insight into math problems and the clarity with which she explained her reasoning to the rest of the class confided in her teacher that she had been "bad at math" in the previous grade, where the curriculum and the teacher had emphasized speed and accuracy in computation. Greta Morine-Dershimer (1983), in a study of six second-, third-, and fourth-grade classrooms, found that the teacher's instructional strategy, and specifically her goals in facilitating full-class discussions, had an important effect on which classmates children listened attentively to and from whom they said they learned.

Depriving All Students of Opportunities to Learn

We tend to worry most about the ways that social ranking affects the lives and learning of children whose classmates have assigned them a low status. They are not, however, the only students who learn less math because of the status hierarchy. Students who have high status frequently miss out on opportunities for powerful learning because they fail to take classmates' divergent ideas seriously. This is a problem partly because the ideas they dismiss could extend their understanding. But another, equally serious problem is that when students fail to think about or entertain ideas that do not immediately strike them as correct, they continue the process of learning math as a list of rules to be memorized rather than as a human construction that engages the mind and involves reasoning, problem solving, communication, and making connections.

Thinking again about how groupwork creates opportunities to learn can help us make sense of this. First, and most obviously, heterogeneous groups provide opportunities for children who are confused about a topic to learn from others who understand it better. Second, children who understand the math better can deepen their understanding by explaining their thinking to someone else. Working in a group gives the child who has no idea how to start on a problem some resources other than the teacher to draw on in his confusion. Since the teacher has many other students to help, the boy who sits waiting his turn to ask the teacher is likely to miss a lot of learning time. If that child has three classmates ready to help him get started, he loses far less time on task. In short, groupwork can deliver direct instruction more efficiently.

Groupwork, when it works well, aids learning in another, far more powerful way: at its best, it forces students to consider more than one way to approach a problem. By the problems we assign, we often teach children without meaning to that doing math in school means looking quickly at a problem and adopting the first solution strategy that comes to mind—usually, the one explained a few pages earlier in the math book. That one is, indeed, what the textbook writer has in mind. To see where this can lead, consider the following problem, given to a large number of children on a test.

There are 125 sheep and 5 dogs in a flock. How old is the shepherd?

Kay Merseth (1993) reports that three out of four American students given this problem produced a numerical response, even though there is no way to get a sensible numerical answer using the information given. Here is one girl thinking aloud as she worked toward an answer:

125 + 5 = 130 ... this is too big, and 125 − 5 = 120 is still too big ..., while 125 divided by 5 = 25. That works! I think the shepherd is 25 years old.

Other children offered similar explanations. These students have learned that they can always solve school math problems by applying a previously taught rule to the numbers provided. Complex instruction aims to disrupt this belief and replace it with the expectation that tasks posed in math class will require thinking. As teachers, we hope that if one student says, "Well, it must be subtract, because if we add the numbers together the answer is too big," another student will raise questions about this reasoning, and the group will work together to make sense of the problem. Ideally, they will realize in the ensuing conversation that they do not have enough information to answer the question. This is unlikely to happen, however, if children have grown used to accepting the answers that particular classmates—those they have labeled "smart"—suggest. When social standing rather than logic decides answers, everyone, regardless of status, learns less.

Here is an example of how this happens from a fourth-grade geometry lesson one of us observed not long ago. The teacher had given each child an envelope that contained a picture of a three-dimensional shape. The students had to identify the shape, count its faces and vertices, and write what they knew about it on their paper (although the teacher allowed children to talk to each other while they worked, she intended this to be an individual rather than a group task).

Jakeel, after examining the picture of a cylinder that he found in his envelope, turned to Nerissa and asked her what his shape was called. "It's a sphere," Nerissa replied confidently.

"S-p-h-e-r-e." As Jakeel was writing *sphere* on his paper, Minerva looked over his shoulder and corrected him.

"That isn't a sphere, it's a cylinder," she said. Jakeel paused, apparently considering the new data, and then finished writing *sphere*.

"Well," he told Minerva, "you are my best friend, but she is smart."

Had Jakeel been less sure that the best path to a correct answer was to follow a "smart" student's directives blindly, he might have asked Minerva, Nerissa, or both why they thought what they did and then evaluated their evidence. Or he might have looked in the dictionary or the math book for information about spheres and cylinders and either learned from them that his shape was a cylinder or talked with Nerissa and Minerva about what he found. Any of these outcomes would have engaged him in investigating and reasoning, activities we want to encourage in math class, and thus created better opportunities for learning than labeling the cylinder a sphere because the student who offered this answer was "smart." (We should, perhaps, note that Nerissa led classmates astray in this way quite often [Wood 2008].) When a group of students labels a particular classmate "smart," they often suspend their critical faculties.

Jakeel is often ready to follow the advice of others he thinks are smart. However, children who have in the past done poorly in math are not the only ones who dismiss correct, potentially educative ideas because of the speaker's status. One spring, Marcy Wood (2010) observed a third-grade class where the children had been exploring ways to divide an area into quarters and halves using 5 × 5 geoboards, where five pegs make up each of five rows, dividing the geoboard into four rows of four squares each.

The children had come to see that they could shape two regions of the board differently but, if the regions' areas were the same, still represent the same fraction of the whole (fig. 2.1). Andy had concluded his work on halves and fourths and, with a little encouragement from an adult, begun to explore smaller fractions. He had discovered that one square in the 5 × 5 board is $1/16$ of the whole geoboard, and that by subdividing that small square he could come up with even smaller fractions—$1/32$, $1/64$, and so on (fig. 2.2).

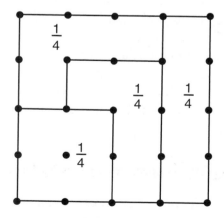

Fig. 2.1. Different ways to show one-fourth on a 5 × 5 geoboard

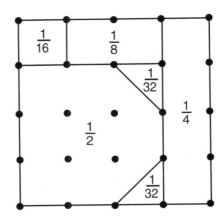

Fig. 2.2. Examples of Andy's fractions

But when Andy came to the overhead projector to present these ideas, his classmates protested angrily that his fractions were wrong, calling out, "I don't think that's right, Mrs. Smyer," "No!" "There's not thirty-two pieces in the whole thing," "But he's not using squares," "He ain't using the whole square," "That ain't right," "No, it's not!" They groaned in chorus and insisted collectively, "But that doesn't make sense."

Although Andy had ideas that could have deepened and extended his classmates' thinking about fractions, and he explained these ideas lucidly, the other children rejected everything he said. The teacher tried to make Andy's ideas accessible to others by asking him questions that would help him present his thinking in a logical and linear way, building on concepts that his classmates already understood. Nevertheless, these third graders, usually polite and well-behaved, on this occasion communicated quite clearly to Andy, their teacher, and the observer that they neither understood nor liked Andy's way of thinking and that they were not interested in learning more about it.

Many of the third graders in this class described Andy as "weird." His family structure was atypical, and he had few friends. As the rest of us watched this episode unfold on video, we saw a vivid illustration of how status differences can limit *all* children's opportunities for learning math. Often, we see status working to limit the opportunities low-status students have to learn by limiting their participation. Here, we saw the reverse: Andy's low social standing prevented his higher-status classmates from taking his mathematical thinking seriously, even though his ideas were more sophisticated than theirs. It seemed that the children did not want to believe that Andy had anything to teach them. Or perhaps his ideas threatened the comforting certainty that math is about applying rules you just learned to fairly straightforward problems.

Reducing the Influence of Status on Learning

Eliminating the influence of status on participation and learning is rarely possible, but complex instruction offers multiple strategies for reducing its effects (Cohen 1997). The stories about Jakeel and Andy help us understand why research shows that both academically successful students and previously unsuccessful ones learn more in classrooms in which teachers use these strategies. Without such strategies—Cohen calls them *status treatments*—teachers and students will continue to reinforce status. Everyone will look to the high-status person to set the work for the group. Even if that person's ideas turn out to be wrong, her taking charge in this instance will elevate her future standing. Similarly, the higher-status children will remember the lower-status ones who did not speak as having contributed nothing to the group effort; the former will expect even less of the latter in the next group assignment (Cohen 1997). But by paying attention to the workings of status in our classrooms and taking steps to disrupt it, we can push our students toward new ways of thinking about math and identifying promising approaches to math questions. We can move them away from dependence on rules and "smart" classmates, toward mathematical reasoning as an approach to solving challenging problems. In chapters 3, 5, and 6, we will look at what teachers can do to change the beliefs about status and math that limit children's chances to learn mathematics.

References

Cohen, Elizabeth G. *Designing Groupwork: Strategies for the Heterogeneous Classroom.* 2nd ed. New York: Teachers College Press, 1994.

———. "Understanding Status Problems: Sources and Consequences." In *Working for Equity in Heterogeneous Classrooms: Sociological Theory in Practice,* edited by Elizabeth G. Cohen and Rachel A. Lotan, pp. 61–76. New York: Teachers College Press, 1997.

Cohen, Elizabeth G., and Rachel A. Lotan. "Raising Expectations for Competence: The Effectiveness of Status Interventions." In *Working for Equity in Heterogeneous Classrooms: Sociological Theory in Practice,* edited by Elizabeth G. Cohen and Rachel A. Lotan, pp. 77–91. New York: Teachers College Press, 1997.

Horn, Ilana. "Learning on the Job: A Situated Account of Teacher Learning in High School Mathematics Departments." *Cognition and Instruction* 23, no. 2 (June 2005): 207–36.

Lotan, Rachel A. "Complex Instruction: An Introduction." In *Working for Equity in Heterogeneous Classrooms: Sociological Theory in Practice,* edited by Elizabeth G. Cohen and Rachel A. Lotan, pp. 15–30. New York: Teachers College Press, 1997.

Merseth, Katherine K. "How Old Is the Shepherd? An Essay about Mathematics Education." *Phi Delta Kappan* 74, no. 7 (March 1993): 548–54.

Morine-Dershimer, Greta. "Instructional Strategy and the Creation of Classroom Status." *American Educational Research Journal* 20, no. 4 (Winter 1983): 645–61.

Rosenholtz, Susan J. "Treating Problems of Academic Status." In *Status, Rewards, and Influence: How Expectations Organize Behavior,* edited by Joseph Berger and Morris Zelditch, p. 445–70. San Francisco: Jossey-Bass, 1985.

Shulman, Judith H., Rachel A. Lotan, and Jennifer A. Whitcomb, eds. *Groupwork in Diverse Classrooms: A Casebook for Educators.* New York: Teachers College Press, 1998.

Tammivaara, Julie Stulac. "The Effects of Task Structure on Beliefs about Competence and Participation in Small Groups." *Sociology of Education* 55, no. 4 (October 1982): 212–22.

Wood, Marcia B. "Mathematizing, Identifying, and Autonomous Learning: Fourth-Grade Students Engage Mathematics." Unpublished doctoral dissertation. Michigan State University, 2008.

———. "Not Understanding Andy: A Metaphorical Analysis of Students' Resistance to Learning." *For the Learning of Mathematics* 30, no. 3 (November 2010): 17–22.

3

Preparing for Groupwork by Teaching Norms and Roles

EQUALIZING status in the classroom creates new learning opportunities for *all* students. It brings students who have in the past had low status into the intellectual work that takes place in groups. Reducing status differences gives these students new opportunities to engage with mathematical ideas. By diminishing automatic deference to high-status students' suggestions, the steps we take to equalize status and broaden participation increase the likelihood that children will actually *think* about the mathematical tasks we assign them. Altering norms, so that high-status students must listen to and engage with the ideas of *all* their classmates, helps these more successful students see problems and mathematical operations from more than one angle. Altering norms thus helps them make mathematical connections that lead to more sophisticated mathematical understandings. This is why high-status students often comment that math is more interesting when they work with classmates on a groupworthy task.

Complex instruction offers teachers a number of ways to address the status and participation issues that inevitably arise in groups. Some of these ways come directly from Elizabeth Cohen's (1994) published work. Math teachers at Railside High School (a pseudonym) developed others as they worked together to address status issues that they saw arising in their math classes. A few come from the student teachers and teachers who have worked with us.

If we are trying to get all our students participating, the most direct approach is to create explicit guidelines that tell them who is supposed to do and say what, along with explicit, consistently enforced norms for groupwork that make equal participation more likely. We can also take steps that help students both see the value of a wider variety of intellectual skills and notice and appreciate contributions made by low-status classmates. Finally, when we need to, we can intervene in groups directly. In this chapter, we'll look at what we can accomplish by giving students well-defined roles to play during groupwork, examine the norms that some teachers establish during complex instruction, and discuss the efforts of teachers we work with to teach new norms and new roles for groupwork. In chapters 5 and 6, we will describe pedagogical moves and strategies that teachers use to support equal participation and mathematics learning during groupwork.

How Roles Affect Status Issues in Classrooms

Students make decisions about who in their class is smart. They often then act on those decisions to give the "smart" students opportunities to participate and lead the group work while assigning the "less smart" students meaningless jobs. This process builds on itself. The "smart" students influence other students' work, and their classmates give them credit for the group's work, thus reinforcing their "smart" image. Meanwhile, the "smart" students, and those who look to them for guidance, exclude students seen as "less smart" from the group's intellectual work. The "less smart" students don't get to try out their ideas or explore the lesson's mathematics. And, finally, the class as a whole prevents the "less smart" students from making contributions to the group, thus reinforcing the "less smart" students' low status. In this vicious circle, the high-status students draw farther and farther ahead of the low-status students, both in status and in math achievement.

In order to interrupt this vicious circle, we assign students formal roles with well-defined duties that draw them into the mathematics at stake in the task. Figure 3.1 shows role cards that teachers and student teachers we know have used with groups. In addition to defining the duties that go with each role, the cards offer sample remarks that children assigned each role might say when enacting their role. This particular set of roles, created by the mathematics departments of Railside High School and its feeder middle school, is, of course, only one of many possible ways to divide up the work a group must do. We continue to adapt the roles and the role cards as we learn more about how younger children interpret them.

Facilitator	Resource Monitor
Gets the team off to quick start	Collects supplies for the team
Makes sure everyone understands the task	Calls the teacher over for a team question
Organizes the team so they can complete the task	Cares for and returns supplies
	Organizes cleanup
"Who knows how to start?"	"I think we need more information here."
"Does everyone get what to do?"	
"I can't get it yet. Can someone help?"	"I'll call the teacher over."
"Can someone explain it another way?"	"We need to clean up. Can you… while I…?"
"We need to keep moving so we can…."	"Do we all have the same question?"

Recorder/Reporter	Team Captain/Includer/Questioner
Gives update statements on team's progress	Encourages participation
Makes sure each team member records the data	Enforces use of norms
Organizes and introduces report	Finds compromises
"We need to keep moving so we can...."	Ensures that each person is doing his or her role
"I'll introduce the report, then...."	Substitutes for absent roles
"Did everyone get that in your notes?"	"Remember, no talking outside the team."
	"Let's find a way to work this out."
	"We need to work on listening to each member of the team."

Fig. 3.1. Role cards created by the mathematics departments of Railside High School and its feeder middle school

Using roles can begin to address status issues by structuring the group's work so that all students, including low-status ones, contribute in intellectually significant ways to the group's mathematical work. For example, the Recorder/Reporter organizes and introduces the group's final product to the class. He *must* understand what the group is doing, so he can fulfill his assigned duties. A group, therefore, cannot brush aside a low-status student in this role and still successfully complete its task according to the established norms.

It is, of course, possible for high-status students to sabotage the roles' intent. For example, some groups may be tempted to have higher-status students create the final product and then coach the low-status student, who is the official Reporter, on a few vague, introductory words. This undermining of the roles will sometimes happen when students are first learning about complex instruction, getting used to rethinking their expectations for themselves and their classmates. It usually diminishes as students become more familiar with roles and norms and, from their teacher's interventions, begin to see new capabilities in themselves and others.

The following report, from one of our student teachers on an elementary school class's first complex instruction lesson, dramatizes both the difficulties of getting started and the potentials of the roles:

> We were really able to see how status affects the students in groupwork. Assigned roles did not seem to change the status of the low-status students.... The high-status students still seemed to dominate the group with directions and opinions, and the low-status students

seemed to fade into the background of the group. It did not occur to us until after the project that the words *team captain* would hold more status than the jobs listed on the cards. We realized this when one of the high-status students responded to a question about his role when he said, "Team captain. Of course, I'm the team captain."

In several instances, the assigned roles did not affect how the classmates perceived low-status students. For example, the team captain at one table basically took over the resource monitor role. He had to be reminded several times that what he was doing was not his job. This happened when the resource monitor went up to the table to gather supplies. The team captain was looking over the shoulder of the resource monitor, telling her what supplies she needed to be picking up. He also told the resource monitor when to raise her hand. When I asked the resource monitor what the question was, the team captain interrupted several times. Although this group had a rocky start to the task, the facilitator got the group together to work as one to solve the problem. She asked questions such as "Does everyone get what we have to do?" and "OK, let's brainstorm what we need." Towards the middle of the task, the facilitator really took the lead and got the group working on an equal level.

As low-status students take advantage of opportunities to contribute to the group's work, they have opportunities to demonstrate their smartnesses to their peers and to themselves. In doing so, they improve their status. One of our student teachers, Eleanor, saw this happen the first time she taught a complex instruction lesson. (We have quoted Eleanor's account at greater length in this book's introduction.)

We followed two [low-status] students in particular—KOD and VIN. The assignment of roles was a big success in pushing KOD higher in status. He was exhibiting a heightened level of confidence through his body language. His chest was puffed up in a dignified position whereas usually he is slumped down sitting outside of the group, doodling. He had the full attention of the group, and he was extremely focused on getting the team's questions just right. His motivation was also improved as, for once, he was not the last one to be done!

This process is an upward spiral: as students participate more, their status improves, which leads to additional participation and increasing status. Susan Harvey, who teaches sixth grade in East Lansing, Michigan, reports that she has been especially pleased with the way this has happened for her students:

Especially our English language learners [ELLs], because they have less confidence in their ability to communicate, they are much more comfortable now getting up in front of the [class]. And they engage in conversations, whereas before they would sit and wait to hear, to have everything wash over them, and not ask questions. They just wanted to keep a low profile. My ELLs have been so much more engaged with what they are doing. And I think that will help them with their language acquisition.

But, of course, things do not always go this smoothly. Eleanor continues her description of her first complex instruction lesson:

Of course, as most things are, this lesson was also not perfect. Although the overall participation of VIN did increase, her social status remained stagnant. Her group members went into the task with their preconceptions of VIN as an incompetent student, and those prejudices affected the way they worked with her negatively. VIN's group members were suspicious that she was giving them inaccurate information by misreading her Build-It cards, when in fact she was reading them correctly all the while. VIN was also physically removed from the group and the manipulatives. But even so, with a little redirection and reminders, her voice was heard sporadically, which is more than a 200% increase in itself.

The roles are designed to ensure that every student participates in the group's work on the mathematical task and that each person's participation is necessary to the task's accomplishment. Some teachers change the role descriptions occasionally to fit the mathematical task's requirements, substituting a Timekeeper for Team Captain/Includer/Questioner, for example, when they are worried about groups staying on task and finishing the assignment in the time available. Dey Ehrlich (Ehrlich and Zack 1997) found that when science teachers enhanced the Reporter's role so that the Reporter encouraged groupmates to use the scientific method, the effect on the group's conversation was dramatic. Ehrlich and Zack also report that some middle school teachers have a "Harmonizer" in each group to keep the peace. In addition, Marcia Zack's research (as reported in Ehrlich and Zack [1997]) indicates that when Facilitators play their roles well, children interact more with one another; the more actively the Facilitator plays his role, the more the children in the group talked about the assignment.

The roles are not supposed to bestow unequal status. However, elementary school students tend to covet particular duties, especially collecting materials and calling over the teacher. Students assigned these roles may believe that they have higher status. Although you may find that you can use a particular role's status to bolster a lower-status student's standing temporarily, roles with unequal status may in the long run reinforce your classroom's status issues. For example, if you consistently assign the Resource Monitor role to low-status students, your high-status students may find ways to give this role lower status. If everyone wants to be the one who calls the teacher over for a group question, the high-status students may sabotage this job's prestige by telling the low-status student to whom you assigned this role when to raise his or her hand and what to say when the teacher responds. Rather than a privileged role of "sole communicator with teacher," the role may become flag person and message bearer.

In order to avoid inadvertently aggravating status issues, one must construct roles that equalize status. This means that, depending on your particular students and context, you may need to change the name of a role or shuffle responsibilities in order to create roles that all students see as important. For example, taking our cue from student teachers who, like the one quoted above, reported that some socially successful students interpreted the title "Team Captain" as ratification of their high status, we now call that role "Includer" or "Questioner." Each role should contribute to the group's intellectual work and function to make group members depend more on classmates in the group and less on the teacher. Thus, although the Resource Monitor is the only person allowed to signal the teacher, she must first check with each group member to make sure that no one can answer the question raised and that everyone has the same question. When the teacher does arrive at the group, he can ensure that the question is truly a group question by asking someone other than the Resource

Monitor what the question is. Resource Monitors, then, are responsible for monitoring the group's use of the teacher as a resource and for ensuring that groups call on the teacher only when they really need to do so.

Finally, one must remember that status arises from perceptions of differences in intellectual competence. Roles, as tools for status intervention, need to provide students opportunities to demonstrate, and be valued for, their intellectual contributions.

The roles in themselves are not enough to change the status of a child whose classmates have for months perceived as incompetent. Roles do, however, give teachers tools for ensuring that all children gain at least minimal access to the task and materials, as well as for deepening the students' engagement with the mathematical ideas. For example, one group of student teachers substituted the title Skeptic for Team Captain/Includer and charged the Skeptic with raising questions about proposed solutions to a problem. One of the Skeptic's lines was, "Wait a minute." This role speaks directly to one of the major problems of group work, one we have mentioned before: groups charged with a common task tend to plunge ahead quite uncritically, deciding in seconds on a line of attack and then dividing up the labor in a way that will lead speedily to the required product. Giving one person the responsibility for stepping back from this headlong rush toward a product and raising sensible questions increases the chance that the group will do the work thoughtfully, with due attention to the mathematical challenges that the task poses. That increased chance, in turn, boosts the likelihood that the students will learn some math. (Of course, a Skeptic can also carry the role too far, paralyzing the group by requiring the equivalent of a legal brief for every suggestion a groupmate offers. Under those circumstances, the teacher may wish to rethink this role or call a meeting to discuss how one can play it responsibly.)

The roles are also not designed to manage unruly behavior. However, children whose classmates have marginalized them sometimes misbehave in order to avoid feeling invisible. When roles help children develop a new awareness of classmates' capabilities, they sometimes open up new ways for the disruptive child to engage with others and with subject matter. Some teachers assign one student the Rover role, which Railside math teachers developed as part of their participation quiz (see chapter 6), when the groups come out uneven. The Rover takes notes on workings of groups and reports to the class at the lesson's end. A teacher–student teacher pair used this role when one of their first graders seemed to have great difficultly sticking to her role, trying instead to do *all* her groups' work. Her teachers gave this little girl the Rover role for one lesson, hoping that she would see what groups looked like when they functioned more cooperatively and be convinced that her classmates could perform their roles. Her job was to observe other groups doing their work, not to interfere at all, and to write down things children did and said that seemed to help the group move forward. The notes this first grader took were excellent, demonstrating eloquently that she had seen other students assisting their groups in ways that did not involve taking over. Her subsequent engagement in groupwork suggested that she had, indeed, learned.

Creating New Norms for Groupwork

Children enter our classes with expectations about what mathematics lessons will look like, sound like, and require from them. Often these ideas differ significantly from the hopes that we as teachers have when we introduce groupwork. Complex instruction teachers create, teach, and enforce norms of interaction to help promote autonomy and interdependence in groups. We want to emphasize that these new norms are intended to foster mathematical learning and not just to manage behavior and status issues.

Here are some we have used during group activities in our university courses:

- No talking outside your group.
- Helping others does not mean giving answers.
- Everyone stays together.
- Ask, "Why?"
- Call the instructor for group questions only.
- You have the responsibility to ask for help and the responsibility to offer it.
- I can't ..., *yet!*

These norms help our students learn to become resources for one another; they also send a message about what doing, and being smart at, mathematics means. The last bullet, a norm developed at Railside, sends a message about how we would like students to see themselves as learners of math. We tell them that if they began a sentence with "I can't..." they have to end it with, "*...yet!*" Joy Oslund reports that by mid-semester, when her students are working in groups, she overhears them stressing the "yets" at the end of their sentences, telling classmates, "I can't solve the problem that way, *yet!*" or "I don't know enough about that algorithm to use it, *yet!*"

Many prospective teachers enter teacher education classes with little confidence in themselves as mathematical doers and learners. Not surprisingly, they are thus afraid to take risks in the seminar. Many have learned in school that mathematics is an individual, competitive endeavor. Some have learned to survive in mathematics courses by hiding behind classmates they perceive as more competent in math, and by staying quiet unless they are certain they have right answers. In the this book's prologue, Lisa Jilk describes doing exactly this in school and college math classes. We chose particular norms for prospective teachers, in order to introduce them to new ways of looking at mathematics: we wanted them to see math as something that one constructs socially and that has as much to do with asking questions as with giving right answers.

Joe Cleary, who teaches fifth-grade math in Holt, Michigan (see chapter 8), calls his norms Groupwork Expectations and posts them prominently in his classroom. His norms are similar, but not identical, to ours:

- No talking outside your group.
- Helping does not mean giving answers.

- No one is done until everyone is done.
- You have the right to ask for help and the responsibility *to* help.
- Follow your group role.
- Call the teacher for group questions.
- Talk and listen equally.
- Show respect to one another.
- Everybody helps clean up.
- No one of us alone is as smart as all of us together!

Although Joe listed and referred to all these norms from the very beginning of the year, he found "No talking outside your group," "No one is done until everyone is done," and "Call the teacher for group questions" most crucial to his students' participation. He stressed them the most as the year progressed.

Different teachers emphasize slightly different sets of norms when teaching with complex instruction. What is important is that we choose norms that help students focus on learning math and we make our expectations for groupwork explicit. This is vital, because these expectations are likely to differ from those students have experienced in the past. Speaking from her long experience as a math teacher, Lisa Jilk urges the importance of keeping the list of norms short, so that teachers and students can focus on and reinforce all of them consistently.

Teaching Roles and Norms

In *Widening the Circle*, Sapon-Shevin (2007) describes two different sets of rules for playing the game Musical Chairs. In the familiar version, ten children compete for nine chairs when the music stops. The person who has no chair is "out" and can no longer play. Each successive round repeats the cycle, with the adult subtracting one chair and leaving one additional child "out," until one "winner" remains. In Cooperative Musical Chairs, an alternative version of the game, all children sit on the chairs when the music stops even though an adult removes one chair before each round. In the first round, the group must accommodate all ten children on nine chairs in order to succeed and move on to the next round; they accomplish this fairly easily by pushing the chairs together and using them as a bench. In the next rounds, all ten children try to sit first on eight chairs, then on seven, and finally on only one. In each round, no one has "won" until all ten children are seated, whether on one another's laps, straddling the cracks between chairs, or by some other means that they invent. As the game progresses, children giggle and squeal as they become a tangled heap balanced precariously on fewer and fewer chairs.

The second version of Musical Chairs has rules different from those of the first, and we would not expect children to know how to play it unless we taught the new rules explicitly. If we gave children these rules, they might question them at first, but, with practice, they could follow them. Although many teachers and student teachers tell us that their students really enjoy playing their roles and doing groupworthy tasks, the teachers who report most success spend substantial time—often, several math periods—helping students work together, learn these new roles, and make the norms real. Sometimes this is easy, and the children take to groupwork like pigs to mud. But often, especially in the upper elementary grades, children have learned other ways to do school through years of practice. If we expect them to work together in particular, unfamiliar ways, we must take time to teach them how to do this and remind them consistently about the new norms throughout the school year.

One of our student teachers described how this played out in the first complex-instruction math lesson she taught:

> This lesson was one of the first group activities [these students] have done. As a result, they had a lot of new concepts introduced to them in one lesson, such as group roles, expectations for group work, and multiple abilities, in addition to a challenging task. In hindsight, I feel that this may have been a lot for them to handle.
>
> For starters, the group I observed overall did not follow their roles. One exception was the resource monitor, Michael. We made it very explicit that only the resource monitor was allowed out of his/her seat, and I think this is the reason why that role was the easiest for the students to follow. They all knew that it was Michael's job to get the materials. However, the other roles (facilitator, recorder, and team manager) were largely ignored. I recorded the time that students finally looked at their role cards: it was 45 minutes into the activity!
>
> In contrast, I was pleasantly surprised at how quickly the students caught on to the expectations for group work, such as "no talking outside your group," "everyone must stay on the same problem," and "your group must agree on a question before asking a teacher." My group needed only a couple of reminders at the start of the task, and after that they were very good about adhering to these expectations.

Joe Cleary, the fifth-grade teacher whose norms we introduced above, also found that he was still explicitly teaching classroom norms in March, after seven months of teaching with complex instruction. One day, he designed a task to deepen his students understanding of equivalent fractions and to teach the norm, "No one is done until everyone is done." The task card and Joe's instructions at the beginning of class required students to check in with their teacher at certain points in their work, to get an ink-stamp approval before proceeding to the next step. About twenty minutes into the lesson, students in one group raised their hands and then quickly lowered them after one boy pointed out, "Hey! Hey, everyone. He is going to make us all explain it to him. Let's practice before we call him over! We have to make sure we all get it!"

Even adults need time to learn new norms. In our university classes, we find that we must continue to teach these new ways of doing math over several seminar meetings. We should expect elementary and middle school students to take longer. When summoned to a group, teachers need to remember to ask, "Is this a group question?" If the group answers, "Yes," teachers can ask a specific person in the group to say what the question is. Often, when teachers ask someone other than the person raising her hand to speak, students assume the teacher has made a mistake. When this happens, we say something like, "This is a group question, so everyone in the group should know what it is. And I'd like to hear about it from Kim."

Insisting that all questions to the teacher be group questions may not sit well with students initially. It may be hard for teachers as well: several have told us that they felt as though they were letting their students down the first few times when they walked away from a child who had a question she had not asked her groupmates. Most report, however, that when they have steeled themselves to insist that they will answer only group questions, children learn to ask their peers before summoning the teacher. The initial difficulties with groupwork norms remind us how differently students are accustomed to "doing school" and how important teaching norms explicitly is. Those reminders help us aid youngsters in understanding not just what we expect of them, but also *why* we expect it and how these new norms support their mathematics learning.

No matter how well we teach the norms, and how well our students seem to learn them, the norms will, from time to time, break down. Remembering her experiences teaching high school math, Lisa writes,

> I found that this happened often when I changed groups or during a particularly stressful time of the year. I would sometimes have to do skill-builders and participation quizzes [see chapter 6] to remind students of the norms and help them to remember that they *did* know how to work together in groups.

In January, recalling how long and hard he had worked to teach his students to call the teacher over only after making sure that no one in their group can answer their question, Joe estimated that in the first months of school, for every ten times a raised hand summoned him to a group, he walked away nine times without giving an answer because some members of the group did not even know what the question was. He answered a true group question one time in ten. His students were accustomed to depending on the teacher for answers and validation of their work. Teaching them to work in a new way, depending more on classmates, took time and energy. But Joe emphasized the importance of teaching students to rely on the resources in their groups by persistently enforcing the "call the teacher for group questions only" and "no talking outside your group" norms. As his students became accustomed to these rules, he noticed a payoff: participation and learning distributed more equally across his students than it had in the past. He has also observed that his students seem to be learning the mathematics content more deeply than in the past. He attributes this difference to the expectation that everyone in a group is responsible for the learning of all that group's members.

Fortunately, good resources for teaching norms and roles exist. Elizabeth Cohen's invaluable *Designing Groupwork* (1994) contains not only excellent explanations of the principles undergirding complex instruction, but also an appendix filled with activities designed to help children learn to work together productively. The Lawrence Hall of Science has also put out some excellent resources (see appendix A).

Our student teachers found it valuable to spend at least one lesson introducing their students to the new norms and roles before engaging them in a challenging, groupworthy math task. Teachers do this in a wide variety of ways, some constrained by the ages of the children involved. Susan Harvey, the sixth-grade teacher we mentioned earlier, introduced her students to complex instruction's roles and norms on the first day of school with an exercise that focused their attention on learning about one another. She started by introducing the following norms.

- Everyone must contribute.
- Respect the contributions of all members.
- Play your role in the group.
- Only talk with the members of your group—no talking to other groups.
- Listen carefully to each member of your group.

She divided the class into groups of four and defined the roles as shown in figure 3.2.

Facilitator
- Make sure your group reads all the way through this task before you begin
- Keep the group together.
- Make sure everyone's ideas are heard.
- Keep track of the time.

Resource Manager
- Collect all materials for task.
- Organize the cleanup.
- Make sure all questions are group questions. (You must all talk about the question and agree that you each have that question before you call on the teacher.)
- You are the only one who can call the teacher to the group to ask the group question.

Recorder/Reporter
- Make sure your poster is organized and complete.
- Encourage your group members to make the relationships among ideas clear on the poster.
- Make sure everyone is able to present your work to the whole class.
- Post the product (your representation of your data) at the front of the room.

Team Captain
- Make sure norms are upheld.
- Keep team focus on task at hand.
- Make sure everyone is fulfilling their role.
- Be the first person to begin sharing information about yourself with the group.

Fig. 3.2. Susan's roles for the Who Are We? task

She gave each group a copy of the task card shown in figure 3.3.

WHO ARE WE? TASK CARD

MATERIALS:
Pencil and paper for each member
Chart paper and markers

PART 1

1. Resource Manager needs to gather materials for the group.

2. Each member of the group should spend two minutes listing things about themselves. This time should be quiet time. (Examples might be: I have two brothers; I love to read; my favorite color is green.)

3. The Team Captain will start sharing with the group.

4. After s/he has spoken, each of the other members will comment on whether or not they share that attribute.

5. Someone will record the information for the group, putting it on a Venn diagram either in the circle for each person or in the center as a shared attribute. See the figure below.

6. Continue around the group sharing information until the teacher asks you to stop.

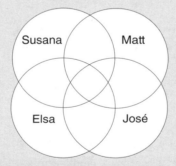

Venn diagram with student names

PART 2

1. When it is time to share with the class, the teacher will tell you whose information you will be sharing. You will not be sharing your own information with the group, so you need to listen carefully to what your teammates are saying and be able to explain what is written in the circle to the rest of the class.

2. Once each member of the team has been introduced, we will move on to the next group.

3. At the end, we'll look at all the intersecting circles to see what kinds of things we all have in common.

WHO ARE WE? TASK CARD—*Continued*

FINAL PRODUCT:

1. Your group's Venn diagram should have information about each person that makes them unique, as well as things that some or all of you have in common with one another. Each team member must have at least four (4) things about themselves represented on your final product.

2. Make sure the representation is readable and neat.

OBJECTIVES OF ACTIVITY

- To learn more about one another
- To practice working in a group
- To follow group norms
- To become learners who can help one another

Fig. 3.3

We think that the private think time in part 1, step 2 of figure 3.3 is an excellent way to get everyone going on the task without creating pressure on anyone to jump in and take over. It also affords each student the opportunity to generate some ideas before the group begins to solve the problem together, and thus may promote broader participation.

The roles Susan used for this first activity were the ones she used all year for groupwork. Teaching the roles took a long time, she reports, but by February she felt confident that her sixth graders could do group assignments even when she was absent and had a substitute teacher in her classroom. Her students knew the roles well and would play them without any reminding. They knew how to cooperate and support one another.

Many teachers, and especially perhaps those in the upper elementary and middle school grades, worry that in an era of high-stakes testing, accelerated and mandated curricula, and accountability, spending time that they could use for academics teaching roles and norms for groupwork is impractical. Susan, however, insists that the time she spent early in the year teaching roles and norms saved her time in the long run, both because her students teach and support one another so well as they work together and because she spends so much less time planning—and replanning—who to put together when she makes up groups.

"And it is so satisfying," she adds, "when I walk around the room and every group is talking about the assigned task."

> I have always been someone who liked groupwork, but the structure of it was elusive. And this just gives you enough of a focus change that you really are concentrating on making every child participate and do something that contributes to the well-being of the group. I don't have to spend the hours I used to spend saying, "OK, I'm going to put this kid with this kid because this kid has this sort of personality and this one has that kind of a personality." It literally takes me five minutes in the morning to shift the groups around.

And it does allow kids to get to know one another. It has been a win-win thing for me. It has saved me a lot of time, but I think it has been real healthy for the kids, too.

The other thing it has changed: I always gave the kids responsibility, but now I give them more. And when they do their presentations, that is when it really comes into play: every kid has to know what's going on. It's not just the leader of the group; it is *everybody*. Now when they get up there they are more comfortable, they are very comfortable, presenting the position of the group. So not only have they personally been more responsible, but they feel the responsibility to the group to do well. And they support one another. And when I am not doing a groupworthy task, they still do the roles. And it is an *amazing* thing to see kids helping one another without being asked to, because they have now internalized a lot of it.

Teachers and student teachers in the primary grades face somewhat different challenges. Many assume at first that complex instruction, the very name of which is intimidating, is not for them and their students. Kindergarten teachers worry that five-year-olds will not be able to handle the social challenges of a group of four. They, along with first-grade colleagues, wonder how they can create useful role cards for youngsters who are just beginning to read. They know that their students will need a lot of help to develop the skills they would need to work together cooperatively.

After taking the plunge, however, many are enthusiastic. Meredith, a student teacher who used complex instruction tasks in the first math lesson she taught her second-graders, wrote:

Our complex instruction day was especially rewarding.... The students maintained a wonderful team atmosphere with their team members and continually told me throughout the activity how much fun it was to work together in that way. They liked the idea that each student brought a unique and needed skill to the table to help them complete their task. They were proud to present their finished products to each other, and the activity served wonderfully as a review for important concepts that the students needed to know for the upcoming assessment.

Summary

In this chapter, we have discussed the norms and roles that support the sorts of changes in groupwork that are basic to complex instruction. We have also talked a bit about how teachers can teach these new norms explicitly, helping their students learn new ways to do math in school. These norms and roles are the foundation of work in groups; they need to be in place if we are to see more equal participation in groups' mathematical work and improved learning for low-status children. But these norms and roles will not, by themselves, change children's beliefs about their own competence or that of particular classmates. For this, we need new mathematical tasks and some different pedagogical strategies. In chapter 4 we examine both the characteristics of a groupworthy mathematical task and the role that a good task can play in equalizing participation. In chapter 5 and 6, we will describe further steps we take to diminish the effects of status on participation as we design and teach lessons.

References

Cohen, Elizabeth G. *Designing Groupwork: Strategies for the Heterogeneous Classroom.* 2nd ed.. New York: Teachers College Press, 1994.

Ehrlich, Dey E., and Marcia B. Zack. "The Power in Playing the Part." In *Working for Equity in Heterogeneous Classrooms: Sociological Theory in Practice,* edited by Elizabeth G. Cohen and Rachel A. Lotan, pp. 44–60. New York: Teachers College Press, 1997.

Sapon-Shevin, Mara. *Widening the Circle: The Power of Inclusive Classrooms.* Boston: Beacon Press, 2007.

4

The Role of Mathematical Big Ideas in Making Groups Work

AS WE WORK to encourage students' more active and more equitable participation in small- and large-group discussions, we must consider carefully the academic task driving those interactions. If we want to offer all our students opportunities to gain experience and fluency talking mathematically, developing and analyzing arguments and evidence, and coming to deeper mathematical understandings, we must present them with math problems that engage important mathematical ideas. It would be impossible to over-emphasize the importance of this point. As teachers, we all have seen how quickly children's conversations wander away from academics when we give them insubstantial tasks. Complex instruction succeeds because, unlike many other approaches to groupwork, it includes a stipulation that the tasks teachers assign to groups be educationally substantive and too challenging for any one student to do alone. Our goal, then, is to base complex-instruction lessons on mathematically rich, intellectually challenging problems. These problems must allow the children to see for themselves the many ways in which they are math-smart. They must also allow us to see students as the thinkers and intellectuals they really are. In this chapter, we begin a conversation, which we pick up again in chapter 7, about design features in a good group task. Here, we focus primarily on the task's mathematics, although we do digress occasionally to point out other features that are too compelling to ignore.

Our first, most important rule when choosing a task is that no student should be able to solve a complex-instruction task alone. A task that one student can do by herself will not impel children to grapple together with big ideas. It will not require conversation, so it will not deepen all participants' mathematical understandings.

Consider, for example, a first-grade lesson on measurement that one of us observed recently. Students sat in groups of four at their small, round tables; their worksheets invited them to choose an object, measure it in centimeters, and record the object's name and its measure ("about _____ centimeters"). The teacher prepared for the task by providing each table with a set of objects—an eraser, a marker, an unsharpened pencil, and so on. She reminded students that they had done a similar task using inches, but that this time they would use the centimeter side of the ruler. The teacher demonstrated at the overhead how to measure one of the objects, stressing the importance of aligning the ruler with the object and starting the measuring at the number zero. She explained that the number on the ruler that lines up with the object's other end tells its length. She asked students to follow the same routine with the other objects.

This particular lesson contains nothing objectionable if its goal is to give students individual practice measuring with a ruler. Neither the worksheet nor this lesson's structure communicated to students that the lesson required them to think and work *together*.

As the lesson continued, several arguments over sharing the objects broke out. At the front table, a little girl grabbed the entire set in order to measure everything before anyone else took a turn. On the other side of the classroom, two students fought over the eraser they both wanted to measure first. At the back table, three boys mocked a child who was carefully measuring and rechecking for being "slow." Additionally, more than half the children started the task by raising their hands to request the teacher's assistance. Such scenes frequently recur when we do not explicitly design tasks as group challenges.

When a teacher uses groupwork to teach mathematics, her goals are quite different from those she might have when assigning individual practice. Instead of focusing primarily on students' independent activity, a teacher using groupwork thinks of learning as a social endeavor. This teacher's goal is to use the structure of groups to support and extend students' mathematics learning. Unless we persuade all children that the given task is worthy and that they will each learn more if they work together, they will have little reason to discuss the mathematics. Also, a problem that is too easy undermines the hard work that the teacher has done to establish the group roles and norms that we discussed in the previous chapter. In this chapter, we focus on how and why to ensure that the complex-instruction task demands group, rather than only individual, effort.

Good tasks are essential to *all* mathematics instruction. In *Professional Standards for Teaching Mathematics,* the National Council of Teachers of Mathematics (NCTM 1991, p. 25) puts the matter succinctly, arguing that worthwhile tasks are ones

> ☐ that do not separate mathematical thinking from mathematical concepts and skills, that capture students' curiosity, and that invite them to speculate and to pursue their hunches. Many such tasks can be approached in more than one interesting and legitimate way;
> ☐ some have more than one reasonable solution.

In order to be worth a group's time, a task must meet these standards.

The work of Stein et al. (2000) helps us think more concretely about what makes a task worthwhile: Stein and her colleagues generated a framework for classifying mathematical tasks according to their "cognitive demand." By cognitive demand, they meant "the kind and level of thinking required of students in order to successfully engage with and solve the task" (p. 11). They identified two kinds of lower-level tasks: *memorization* and *procedures without connections.* Teachers usually give these to students as a set of 10–30 similar problems to solve at one sitting. For example, a worksheet that has students practice converting fractions to decimals and percents may require very little more than marching step by step through a procedure multiple times. In contrast, higher-level tasks—*procedures with connections* and *doing mathematics*—invite students to figure out relationships and build connections they have not already learned. One such task might invite students to explore relationships among various ways of representing fractional quantities. Of course, different learning goals require different tasks. Stein and her colleagues (2000, p. 14) mention groupwork goals specifically:

> Not all tasks provide the same opportunities for student learning. Some tasks have the potential to engage students in complex forms of thinking and reasoning while others focus on memorization or the use of rules or procedures. In our work with teachers ..., we discovered the importance of matching tasks with goals for student learning. Take for example the case of Mr. Johnson (Silver and Smith 1996). Mr. Johnson wanted his students to learn to work collaboratively, to discuss alternative approaches to solving tasks, and to justify their solutions. However, the tasks he tended to use (e.g., expressing ratios such as $^{15}/_{25}$ in lowest terms) provide little, if any, opportunity, for collaboration, exploration of multiple solution strategies, or meaningful justifications. Not surprisingly, class discussions were not very rich or enlightening (p. 14).

Students will not naturally organize themselves for collaboration when solving a math problem unless the given task challenges their individual math smarts and suggests to them that, by working with others, they will have a better chance not only of conquering the challenge but also of learning more. Elizabeth Cohen's (1994) discussion of the essential features of complex instruction tasks echoes this concern, as well as other ideas that math educators have considered important for many years. She writes (p. 68),

> When objectives are conceptual rather than routine, you will want to find or create a multiple ability task: a task [that requires] a wider range of intellectual abilities than conventional school tasks.

Unlike most of those who write about the qualities of good school mathematics tasks, Cohen emphasizes that some tasks that will not support students' learning in groups may be worth assigning to individuals, either as homework or seatwork. Telltale signs that a task will not work well as a group assignment include the following.

- The task has a single, right answer.
- One person can do the task more quickly and efficiently than a group.
- The task's cognitive demand is low.
- The task involves simple recall of memorized material or practice of known routines.

A Groupworthy Linear-Measurement Task

The previous discussion helps us see that the "measure with a ruler" lesson that we described earlier focuses on too narrow a set of skills to work well as a group assignment. Many measurement tasks, however, *are* good candidates for groupwork. After all, measurements are never exact and thus do not yield a single, right answer. Furthermore, several people can often do the practical, physical aspects of measuring more easily than an individual.

Depending on the setup, measurement tasks can provide much occasion for mathematically productive discussion among group members, who must convince one another that their work is trustworthy. In fact, NCTM's *Principles and Standards* (NCTM 2000, p. 173) contains such a task (see fig. 4.1).

Measure and compare:

Work in pairs and use your rulers to measure the items indicated on the chart. Record your measurements for each object on the chart.

	Objects		
	Height of Teacher's Desk	Circumference of Clockface	Length of Classroom
Jo and Rustin	70 cm	92 cm	8.0 m
Whitney and Beth	68 cm	96 cm	7.5 m
Ben and Anna	61 cm	91 cm	8.2 m

Fig. 4.1. Sample linear-measurement task for grades 3–5 (NCTM 2000, p. 173)

The NCTM (2000) task reproduced in figure 4.1, like many such tasks, sends students to measure different objects' lengths using a measuring instrument—in this instance, a ruler. The authors of *Principles and Standards* offer it as an example of a task that provides students with opportunities to "gain facility in expressing measurements in equivalent forms," "use their knowledge of relationships between units and their understanding of multiplicative situations to make conversions," and "encounter the notion that measurements in the real world are approximate" (NCTM 2000, p. 172). Since one goal of this particular linear-measurement task is to learn that different people will come up with different measures for the same object even when using the same measuring instrument, it makes sense to assign the task to a pair or a group, rather than to an individual.

By sending students to measure three challenging objects (their measures are longer than a ruler's typical length, and one of the lengths is circular), the task helps problematize the measuring process. In doing so, it focuses students on important mathematical issues. Once all the pairs of students have shared their measures, we can imagine the class having a discussion about the discrepancies in their measurements and how to resolve them. This allows the teacher to introduce the idea that all measurements are approximate and to push students to think about when, and how much, precision is necessary. Pedagogically, the task sends the children to measure in pairs, thus distributing the work and increasing the likelihood that the students will discuss the problems while they work. With luck, the students' answers will become part of the math lesson, contributing perhaps to the students' interest in, and feeling of ownership of, the lesson.

Similarly, the complex-instruction version of the linear-measurement task that we share next (see fig. 4.2) illustrates how a good group task requires students to contend with the challenging ideas related to meanings and processes of measurement. As you read through the task in figure 4.2, notice the similarities and differences between this task and NCTM's Measure and Compare task in figure 4.1. We will focus next on differences between the two tasks, in the mathematical work we send the students out to perform, the math smarts each task requires, and how the tasks encourage students to work together.

Linear Measurement Task Card

by Mary Beth Blake and Marcy Wood

Materials:

4 rulers

8 sentence strips or pieces of tag board

Scissors

Objective of Activity:

As a group, make a paper strip that shows the length of each measurement below. The goal is to get smarter about strategies we use when measuring linear lengths. What strategies do we use when measuring with a ruler, or without a ruler? How does each strategy handle accuracy?

Task:

Part 1

Each person completes the following steps.

1. Choose one measurement from the list below.
    (No two people should have the same measurement.)

    23 centimeters

    5 inches

    9 inches

    8 centimeters

2. Measure and mark the length on the paper strip.

Come back together as a group.

1. Take turns presenting your paper strip and its measurement to the group.

2. As a group, determine if each measurement is correct, and help the measurer make changes, if necessary. The person who made the measurement should make any changes.

3. When the group agrees that the measurement is correct, the measurer should cut the strip at the mark, and each person should sign the paper strip.

Linear Measurement Task Card—*Continued*

Part 2

Repeat the steps above using the following measurements:

- 17 inches
- 2 feet
- 62 centimeters
- Half a meter

Part 3

As a group, make a poster that uses words, pictures, and numbers to show one strategy that you used to make one of the measurements in part 2.

Extension:

For each measurement in parts 1 and 2 above, find a body part that is the same length. You should find at least two measurements on each person. Each person should make and label a picture showing these two measurements.

Fig. 4.2. Linear-measurement complex-instruction task

Mathematically, the NCTM and complex-instruction linear-measurement tasks above are quite similar: both have students measure things, and in both, the resulting measures become the object of discussion about the measuring process. Both tasks suggest that measuring requires multiple smartnesses: using a measuring instrument, expecting and resolving measurement errors, comparing lengths, comparing units of measurement, and developing strategies for measuring lengths that are longer than the measuring instrument.

An important difference between the two tasks is that the complex-instruction task has students *construct* a length with a given measurement, as opposed to finding a particular object's length or matching a given length to a particular object. In the act of construction, a measurement becomes something that one can create rather than something that just *is*. This act of creation heightens students' sense of ownership and contribution to the math lesson. They have generated not only a data point for the lesson, but also an object that has their name on it. Furthermore, constructing and building are smarts that many students bring to school, yet we seldom recognize their value in the mathematics classroom. This complex-instruction task provides an opportunity for students with building experiences to participate meaningfully in the lesson's intellectual work and be recognized as making an important mathematical contribution.

Another important mathematical feature of the complex-instruction task is that it engages students with two different systems of measurement. It requires them to work with inches and feet, centimeters and meters, and whole numbers and fractions. It has them explain and justify their measuring strategy and discuss and make sense of different units of measurement. In the words of Elizabeth Cohen, the complex-instruction version of the linear measurement task is a *multiple-ability task*. It provides more entry points and pathways through the task, and it therefore offers more opportunities for students with different talents to contribute and for children to need one another as intellectual resources.

Pedagogically, the NCTM and complex instruction tasks (figs. 4.1 and 4.2) are similar in that they send students to work together. The NCTM task, however, leaves unclear what, if anything, will prevent one of each pair from doing all the work while the other simply watches, holds the ruler, or records answers. In fact, one of us recently observed sixth graders working in pairs on a similar task. We saw boys and girls who were paired together adopt stereotypically gendered roles: the boys did the work of measuring while the girls acted as secretaries.

A final difference between the two tasks is that the NCTM task leaves it to the students to figure out how to distribute the work and ensure that each person in the pair or group participates, contributes, and learns. In contrast, the complex-instruction task explicitly directs students to discuss their measures in their groups, not just to wait for public sharing later when the whole class gathers to debrief. It warns students to expect discrepancy in their measures and gives them a process for resolving those discrepancies, potentially preventing one group member from taking over and fixing all the questionable measures. It supports developing the group norms of individual and group accountability by requiring students to check one another's work and take responsibility for their own and everyone else's work.

When we designed the complex-instruction task, we believed that it had many features that would encourage students to work together. As we've used the task with different groups, we are consistently surprised by how well students, even those who have had little or no work on norms and roles, collaborate on the task. In the next section, we describe some of our work with groups and what those groups have taught us about this task's groupworthiness.

Students Work on the Complex-Instruction Length-Measurement Task

When we used this complex-instruction task with third-grade students, we found that the children worked enthusiastically, claiming ownership for one of the four lengths their group had to construct, marking their measurements on their strips, and presenting their work to their peers. Although the students had prior experiences with measurement, many had difficulties using the rulers to measure lengths accurately. We noticed this especially during the second part of the task, when the target lengths were longer than their 12-inch (30 cm) rulers, necessitating several iterations of the ruler.

Many students started measuring at the very edge of the ruler instead of at 0 and had a hard time deciding when they had the appropriate length. Matt, for example, seemed to be struggling to measure 17 inches. When the teacher asked him to explain his thinking, Matt said that he was not sure how he was going to measure 17 inches when his ruler ended at 12. Rather than offer an explanation, the teacher asked if someone else in Matt's group thought they could help. Several students volunteered. Before stepping back to allow Matt's group to support his work, the teacher told the group that she would be back shortly and when she returned, she wanted Matt to be able to explain how to measure 17 inches. As the teacher watched from a distance, Matt's peers picked up the ruler and began talking to him about how to measure. After several moments of conversation, in which Matt asked questions and worked with the ruler, he constructed a measurement. When the teacher returned to the group a little while later, he showed her how he knew that the length was 17 inches.

This story illustrates some of a groupworthy task's important aspects. The students took the certification portion of the task seriously: no one student did his or her work alone. Thus, when Matt needed help figuring out how to measure, the other children found it more important to be sure that Matt could complete and explain his work than to race to finish their own measurements. Also, because each student had a different measurement, they each faced different mathematical challenges. When it came time for the students to certify one another's work, they had to make sense of a measurement they had not worked on, which required both a clear explanation from the presenter and the attention of everyone else in the group. The different measurements and the certification piece seemed to provide a challenge for all students while also requiring that they be accountable to, and interdependent on, one another.

The next time we tried the task with elementary school students, we changed it slightly (see fig. 4.3). This time, students had to construct the lengths called for in part 2 without the aid of rulers. Instead, they had to use the paper strips that they had constructed in part 1 as measuring instruments. This variation of the task heightens the importance of the first set of measurements, since future measures depend on the first set's accuracy. It also helps equalize status in the group, because everyone's paper strip plays a crucial role in the group's work on the part 2 measurements.

Part 2

Use the paper-strip measurements your group created in part 1 to construct the following lengths. Use the procedures you used in part 1 to check your measurements. *Do not use rulers except to check your results afterward.*

- 7 inches
- 2 feet
- 62 centimeters
- Half a meter

Fig. 4.3. Variation to part 2 of the task in figure 4.2

Because we wanted to see whether this task could ensure that students worked together without anyone taking over and doing all the work, we tried it with a group of children who had not yet worked together on an academic task. We suspected that collaborative work would be difficult to achieve without an adult's heavy hand. We were surprised and delighted to see that the children cooperated without much prompting, especially in constructing the measurements for part 2. At first, the children worked independently on this part of the task, each child attempting to use the strip he or she had made in part 1 to construct the strip for part 2. However, they soon realized that they needed to combine one another's strips to build the strips for part 2. For example, using the five-inch and nine-inch strip together would give them 14 inches, which they could fold in half to make the seven-inch paper strip they needed in part 2. In fact, the children figured out that they needed two steps to make 62 centimeters. First, they used the 23-cm strip twice (23 + 23 = 46 cm) to make 46 cm, to which they added the eight-cm strip twice (46 + 8 + 8 = 62 cm) to make the 62-cm paper strip. Our observations indicated that the task's structure was quite effective in getting the children to work— and think—together on challenging mathematics.

This story also illustrates how a small change in a task—in this instance, requiring students to use the measurements from part 1 to construct measurements in part 2—can both improve collaboration among the students and add important mathematical challenges to the task. As students combined their strips in different ways, they needed to consider composing and decomposing numbers and using part-whole relationships among numbers, two skills central to much of elementary school mathematics. The significant increase in group-worthiness arising from this small change impressed us. You will see something similar happen in chapter 7.

Teachers' Work on the Length-Measurement Task

We have also used this task with classroom teachers and student teachers. We want to mention briefly some of our experiences with these adults, because they raise mathematical issues that you might want to consider in your work with elementary school students.

Accuracy

The teachers' groups we work with frequently engage in debates over the accuracy of their measurements. In one instance, a student teacher teased a groupmate, saying, "Aren't you over by 2 mm?"

The groupmate then rolled her eyes and retorted, "That's pretty close!"

This issue of accuracy brings the group to discuss nontrivial issues associated with how they are measuring their paper strips. How are they aligning the ruler with the paper strip? Are they starting at the edge of the ruler or at zero? How are they drawing the line where they are going to cut the strip? Is that line "straight?" Are they measuring both edges of the paper? These are important concerns for elementary school children as well. We think you could raise them with your students by asking questions about how accurate measurements need to be.

How Close Is Close Enough?

Occasionally, one group of adults will finish well before the others. In checking in with the early finishers, we have almost always found that they have used estimation strategies to find each length in the task's second part. For example, one group used their knuckles to add two inches to the five-inch measure to get seven inches. They employed a similar strategy to subtract two inches from the five-inch measure in order to add it to the nine-inch measure to get a 12-inch length, which they then doubled to make two feet. When other groups challenge these estimation strategies, discussions usually bring out the idea that all measurements are estimations because we could always make our tools more precise—for instance, by going from centimeters to millimeters.

A note about classroom management: although discussing estimation strategies can be extremely productive, we usually ask groups that have used these estimates, instead of combining the strips from part 1, to figure out ways to find more precise measures for at least two of the lengths, in order to ensure that all groups finish at approximately the same time.

How Many Possible Lengths Can We Create?

Groups often realize that generating a greater variety of lengths will give them more flexibility. Some fold the strips from part 1 into halves and fourths to get smaller lengths. Others combine longer and smaller strips to find the missing part—for example, using the 23-centimeter and eight-centimeter lengths to create a 15-centimeter length. Many groups are thrilled when they figure out a combination that creates a one-inch or one-centimeter length: This, they realize, enables them to find any length they need. For example, they can do one-inch measures by subtracting the five-inch strip from the nine-inch strip to create a four-inch length and then subtracting the four-inch strip from the five-inch strip. For metric lengths, they can create a 15-centimeter strip by subtracting eight centimeters from 23 centimeters. Subtracting eight centimeters from 15 creates a seven-centimeter length; subtracting that seven-centimeter length from the eight-centimeter length creates a one-centimeter strip. We believe that the challenge of creating one-inch and one-centimeter lengths would be an excellent extension activity for elementary school students. In addition to promoting flexible thinking about numbers, it also would encourage conversations about what makes an efficient measuring tool.

We watched this task transform one student's mathematical participation. A young woman who had previously identified herself as weak in mathematics came up with the idea of folding the five-inch and nine-inch strips in half in order to make the seven-inch strip. Her group seized on this idea as "brilliant," which fundamentally changed the way this young woman worked with others throughout the rest of the course.

Summary

We have tried here to introduce principles central to selecting or designing a good math task for a complex-instruction lesson, emphasizing especially that mathematical ideas are crucial to the success of a math lesson. Children will learn what we give them an opportunity to think and talk about. We will dig into these ideas more deeply in chapter 7, "Selecting and Designing Groupworthy Tasks," and in Appendix B. But before we explore task design further, we need to introduce some of the pedagogical strategies that complex instruction offers a teacher. These strategies and instructional moves can help a teacher reduce the effects of status on students' participation, and hence on learning. Understanding these elements of complex instruction will make what we have to say in chapter 7 about task design easier to follow.

As important as the groupworthy task is to designing a lesson that will deepen students' understanding of important mathematical ideas, a teacher needs more than a challenging, groupworthy mathematical task to create a math lesson in which all students participate in the intellectual work and learn mathematics. He must also build into the lesson pedagogical moves that sensitize the children to the potential contributions of all their classmates, especially to those of children who have in the past been very quiet or unsuccessful in math class. In addition, the teacher needs to create some structures that ensure that the children will be accountable not only for their group's "product," but also for their own learning. In the next chapter we take up creating what Cohen and others call a *multiple-abilities treatment,* as well as structures for individual and group accountability.

References

Cohen, Elizabeth G. *Designing Groupwork: Strategies for the Heterogeneous Classroom.* 2nd ed. New York: Teachers College Press, 1994.

National Council of Teachers of Mathematics (NCTM). *Professional Standards for Teaching Mathematics.* Reston, Va.: NCTM, 1991.

————. *Principles and Standards for School Mathematics.* Reston, Va.: NCTM, 2000.

Silver, Edward A., and Margaret Schwan Smith. "Building Discourse Communities in Mathematics Classrooms: A Challenging but Worthwhile Journey." In *Communication in Mathematics, K–12 and Beyond,* 1996 Yearbook of the National Council of Teachers of Mathematics, edited by Portia C. Elliott, pp. 20–28. Reston, Va.: NCTM, 1996.

Stein, Mary Kay, Margaret Schwan Smith, Marjorie A. Henningsen, and Edward A. Silver. *Exploring Cognitively Challenging Mathematical Tasks: A Casebook for Teacher Development.* New York: Teachers College Press, 2000.

5

Addressing Status Issues through Lesson Design

NORMS AND ROLES are essential for changing students' participation. However, they will not, by themselves, result in the equitable participation we desire. We must also design lessons carefully and intervene as lessons are under way, in order to support and guide students toward engagement with one another and rigorous mathematics. The next two chapters describe how to do this. Chapter 5 focuses on things that we do when we design a math lesson—create a multiple abilities treatment, and build both individual and group accountability into the task. In chapter 6, we consider some instructional moves that teachers can make while the students are working in their groups.

Multiple Abilities Treatments

The term *multiple abilities treatment* sounds much more clinical than it is. Simply put, the "treatment" is a matter of launching a lesson by listing many different academic skills and abilities students will need in order to succeed at the task. The teacher follows this list with the explicit message that no one student has all the abilities necessary to complete the task successfully, but that a group of students, together, will have the skills they need to succeed. Of course, in order for the treatment to be effective, the task must actually require many different abilities. We will explore how a multiple abilities treatment addresses status and then provide examples of two such treatments, but first a few words about the term itself.

We use the term *multiple abilities treatment* here because others working with complex instruction have used it and we want to give the reader access to their work. We do, however, have two concerns about it. First, we believe that many people use the word *ability* to refer to fixed, even innate, capacities rather than learned skills and proclivities. Second, the word *treatment* seems to suggest something that one does once to cure some sort of malady, whereas you would probably do a multiple abilities treatment many times. If you share our reservations about *multiple abilities treatment,* you might prefer *orientation to multiple smartnesses* or *talent and skill inventory*—or another term we have not thought of.

Designing multiple abilities treatments forces us to be specific and explicit in our effort to change students' ideas about what excelling at math means and, in doing so, creates possibilities for changing status relationships. As we explained in chapter 1, the narrower our ideas about what it takes to be good at math, the fewer children will see themselves—and be seen by others—as competent math students, and the fewer chances most children will have to succeed. Two things drive our efforts to broaden students' ideas about being good at math. The first is our conception of what math is and how one comes to understand it. The second is our determination to change the way small groups work: we want to encourage students to listen with genuine interest, and not mere politeness, to *all* their classmates' ideas and to try to use these ideas in looking for solutions to math problems. When children listen seriously and try out one another's ideas, everyone benefits, because all encounter new ways to reason mathematically.

A teacher creates a multiple abilities treatment after, or occasionally at the same time as, she designs the math task, but before teaching the lesson. She does this by looking at the problem and considering what sorts of skills, or "smarts," students might use to work on it productively. Helen Featherstone designed a task for student teachers in a teacher education class that we will use to illustrate the thinking we do when we create a multiple abilities treatment. She constructed this task to help prospective teachers think about the difficulties elementary school students may have in understanding area and metric conversion and about how to plan a discussion that can support children's math learning. We have copied Helen's task card below (fig. 5.1), but because Helen designed the task for college students rather than elementary school children, you may want to skim the task card and then go directly to the section heading Multiple Abilities Treatment for "Discussing Area," below. You can do this without missing anything important to your understanding of the treatment. We include the task card here because it shows in some detail how Helen thought about creating the treatment.

First, the task card:

Discussing Area

As a group, plan a discussion for the conclusion of a sixth-grade problem-solving lesson.

1. Materials Managers hand out a shape piece (see fig. 5.2) and a white card to each person.

2. Each group member works alone for two minutes to figure out the shape's area in square centimeters and square millimeters. Each group member writes his or her thinking on a white card and gives the card to the Materials Manager, who puts it in the group's envelope, labels the envelope with the group's number, and takes it to the resource table.

3. Each person shares his or her thinking on the shape's area. Each person speaks without interruption the first time. After everyone has had a chance to share their solution and their reasoning, the group tries to reach consensus on the shape's area in square centimeters and square millimeters.

Discussing Area—*Continued*

4. When the group reaches a consensus, the Materials Manager gets the envelope of the group whose number is one higher than theirs from the resource table. The group with the highest number—say, #6 of six groups—gets group 1's envelope. Together, all group members examine the solutions in the envelope they received to see if these solutions do the following:

   a. Challenge the group's thinking, or

   b. Offer them insights into how others might think about the problem.

5. Now your group should plan your discussion. This is the background you will need in order to plan:

   > Your sixth-grade class has been working on measurement for several weeks. The children have spent time on inches, feet, miles, yards, and metric linear measurements—centimeters, millimeters, kilometers, and meters. They have also done work in area, mostly in square inches and square feet.

   a. Writing your discussion plan:

      i. Analyze the problem (the shape's area in fig. 5.2), its potential, and so on

         1. Make a *list of difficulties that children may have* in solving the problem, and explain why you think they will have them.

         2. Make a list of *mathematical misconceptions* children may have that you hope the discussion will uncover and shed light on.

         3. List your *goals for the discussion*.

      ii. Plan the moves you will make *during* class in order to make a good discussion

         1. Explain what you can do *during the work period* that will help you get the children involved with the math and with one another's ideas later, during the discussion.

         2. Explain how you will *launch* the discussion in order to get the children involved with the math and with one another's ideas.

         3. What moves will you make *during the discussion period* in order to get the children to listen to one another and think about and respond to one another's ideas?

   b. Write your plan on paper, and tape it to the blackboard.

Fig 5.1

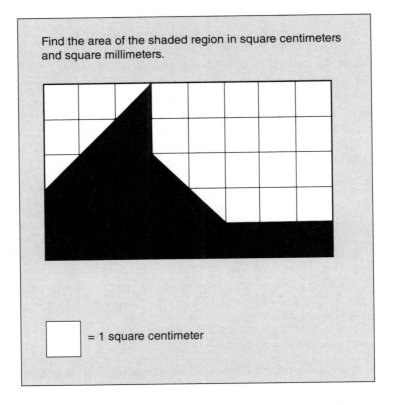

Fig. 5.2. A figure reproduced from Stein (2001, p. 110)

Like most groupworthy tasks, Helen designed this one with multiple goals. She wanted the student teachers to do the following:

- Think about what converting square centimeters to square millimeters involves and how relevant or irrelevant the algorithms for computing area are to an irregularly shaped figure
- Learn more about planning a math discussion that would engage children in constructing new understandings
- Analyze a task, focusing on what sorts of difficulties children might have in doing it
- Consider some possible student errors, using both their own errors and the evidence from the envelope about fellow students' very first, and sometimes mistaken, thoughts.

Helen also wanted to provide the student teachers a space for thinking together about how to use this information in planning for a discussion.

Having designed the task, Helen listed for herself some of the skills that students might use to help their group succeed. The first four items on the list relate to the math task; the others pertain to the pedagogical task, planning a good full-class discussion. Here she explains the list.

Careful Reading

Ninety percent of solving many difficult math problems is figuring out what the problem really asks. This is not obvious to students who have had only limited success in math. Many are relieved to learn that their reading skills, in which they have considerably more confidence, can help them with math. Careful reading would help students identify aspects of this task that they might otherwise miss.

Time Management

A group probably will not finish a multistep task like this one in the time allotted unless they create and follow a plan for using their time well. That shouldn't mean shushing anyone who raises a question about a proposed solution. It should mean reading the directions through before jumping in, and then keeping one eye on the clock and the other on the jobs to be done. If group members do not plan their time, they will end up rushing to produce some sort of product in the last few minutes of the work period, in their haste valuing words on the page more than thoughtful discussion.

Understanding Children's Ideas about Area

Most people have had far more experience with linear measurement than with area measurement. This task requires the student teachers to think about what it means to measure area.

Spatial Thinking and Geometric Imagination

The math problem requires the students to think spatially about the relationship between square centimeters and square millimeters and how this might differ from the relationship between centimeters and millimeters. The problem also requires that students find a way to compute area that does not depend on applying a formula.

Pedagogical Imagination

The task requires the student teachers to imagine multiple ways that sixth graders, either individually or in groups, might respond to the problems posed. This kind of pedagogical imagination does not come readily to prospective teachers who have still spent far more time as students than as teachers. As students, they saw many math teachers in action but got few opportunities to compare a lesson's actual unfolding with the ideas and intentions of the teacher who planned it. Experience has not yet taught them how many different ways a lesson can go. A student teacher who has the patience, talent, and inclination to imagine multiple scenarios can help his group make a sturdy lesson plan that will continue to provide guidance even after the students have voiced unexpected ideas.

Flexible Thinking

Both the math and the pedagogical tasks require the student teachers to revise a plan already made in light of new information: when they open another group's envelope, they get new data about how others have thought about the area of the shape— data that they may be able to use to rethink their solution to the area problem and to anticipate difficulties sixth graders may have with the task. This new information may permit the student teachers to improve their plan, if they can think about it flexibly.

Curiosity

A student teacher who is curious about the ideas of those whose thinking about a math problem differs from his own has a far better chance to see a problem from multiple angles, to learn about the difficulties that constrain children when they struggle with a problem like this one, and to learn from the opportunity to plan collaboratively.

Helen then constructed a multiple abilities treatment by writing this list (fig. 5.3) on a transparency and putting it up on the screen before she distributed the task.

Reading skills

Time management

Understanding area

Spatial thinking

Imagining multiple scenarios

Geometric imagination

Flexible thinking

Curiosity

Pedagogical imagination

Fig. 5.3

Multiple Abilities Treatment for "Discussing Area"

In presenting the multiple abilities treatment to her students, Helen said something like, "These are skills that your group will need in order to complete this task. No one person will have all of these skills, but all of you have some of them to contribute to the success of your group" (see Cohen [1994], p. 128 for a discussion of this idea). Helen did not want her students to use the multiple abilities treatment as a checklist of activities and skills to tick off as they worked, so she took the list down once she had given the students a chance to look at it and ask any questions that it triggered for them.

In creating this list, Helen had the following five goals.

Goal 1: Assessing the Task's Groupworthiness

The first audience for Helen's task card and her multiple abilities treatment is Helen herself: writing a multiple abilities treatment requires her to look closely at her task, to see whether it will call forth a variety of mathematical skills and smartnesses. If she struggles to find things to write on the multiple abilities list, she knows that the task is not as multidimensional as it ought to be. If she thinks that the task would engage her students in some worthwhile thinking and help them hone important skills, she might still go ahead with it, but she would know that some time soon she will need to design or find a problem more multidimensional and more likely to call forth some skills that go beyond the conventional ones.

The multiple abilities treatment does not simply help students see the smartnesses of a wider range of classmates; it also offers them more than one route into and through the mathematical ideas. Giving students access to multiple routes provides a chance to forge new connections between mathematical ideas they already understand and those ideas that they are encountering for the first time. Writing a multiple abilities treatment is a way for a teacher to assess the potential groupworthiness of a math task.

Goal 2: Drawing Attention to Parts of the Task That Students Might Otherwise Ignore

Often, groups will rush ahead when they see a long, or even short, assignment, putting down the first thing someone suggests, determined just to get done. Skill sheets that require students to practice a particular algorithm over and over reward this approach. When Helen included *reading skills* on the skills list, she was trying to suggest to the student teachers that (a) she wanted them to read carefully, because aspects of this assignment might be less straightforward than they appeared; and (b) that they should pay attention to any group member who said something like, "Wait, I think we missed this part…." Similarly, when she wrote *imagining multiple scenarios,* she signaled that she expected the student teachers to try to identify more than one way that the lesson could unfold—a difficult task for beginning student teachers, who do not yet have much classroom experience to draw on—and to consider all these possible scenarios when making their plans. She has also tried to suggest these things in the task card itself, but she knows from experience that directions on a task card will not by themselves change students' longstanding habits. Neither will a list on the overhead projector, but the more ways she can support students' impulse to stop and consider multiple outcomes, the better for their future planning and teaching.

Goal 3: Encouraging All Students to Believe That They and All Their Groupmates Have Important Intellectual Skills to Contribute to the Work at Hand

By calling attention to the range of skills that the group would find useful in working on this task, Helen hoped to convince the prospective teachers that no one of them had all the skills that the task required and that they could not do the task well without a deliberative conversation that elicited everyone's ideas. She wanted students who usually hung back during math discussions to find skills on the list that they knew they had and gain confidence that they could contribute to the success of their group. She also hoped to focus *all* students' attention on the other group members' potential contributions, in order to enhance the status of students whose skills they may have often been overlooked in the past. Until children come to believe that many legitimate ways exist to approach a math problem and that listening to ideas that at first seem strange can lead them to new mathematical understandings, they will not learn everything they can from groupwork.

Goal 4: Expanding the Student Teachers' Ideas about Skills and Dispositions That Contribute to Mathematical Competence

The multiple-abilities treatment offers us a concrete way to work on expanding students' ideas about what math is, what doing math involves, and what sorts of smarts help a person learn math. The more successful we are in enlarging students' ideas of what goes into the bag labeled *math smarts*, the more opportunities we will have to convince them that many students are smart in math and are therefore worth listening to.

Notice that these goals are linked: unless students notice the task's breadth and complexity, they will probably not appreciate contributions that depend on skills beyond computing area. Unless the teacher designs a task that uses multiple skills and strengths, she will not create ways for all students to demonstrate their mathematical and pedagogical smarts. By creating a venue where students will need to use a variety of intellectual skills and strengths, and by calling attention to the importance of these contributions, the teacher can disrupt the existing status hierarchy and improve the chances that students will hear and appreciate the contributions of more of their classmates.

We present the multiple abilities treatment at the beginning of the lesson, as we introduce the task. We can add considerably to its power by taking note, while the children work on the assignment, of moments when a student uses one of these abilities. By pointing out these occurrences to the student and his classmates, we highlight for students how their multiple abilities advance their work and deepen their mathematical understandings. When a teacher launches a math lesson with a multiple abilities treatment, she sets herself up to look for, assess, push, and highlight publicly students' use of the different skills she has named. This is the accountability piece. Telling students that they will need these different skills is not enough. The teacher must point the skills out and hold students accountable for actually using them. We will say more about this in chapter 6.

Our second example of a multiple abilities treatment derives from a sixth-grade math lesson. Teresa, a student teacher, developed "It's Cooking Time" to support her students' work on fractions. At our request, she wrote a bit about the context:

☐   Each week I try to formulate a groupworthy activity in which all of my students will become engaged with the topics we are currently studying. When I developed "It's Cooking Time," my students were learning how to add fractions with uncommon denominators. I wanted to create a groupworthy activity that had connections to real-world scenarios as they were dealing with fractions. Baking is one activity that frequently requires the use of fractions, so I knew I wanted to utilize it in the groupworthy task.

    Adding fractions was certainly the main objective of this activity, but I also wanted my students to review other ideas such as converting decimals to fractions and vice versa, breaking whole numbers down into fractions, and converting improper fractions to whole
☐   numbers.

Figure 5.4 shows the six clue cards, and figure 5.5 the task card, that Teresa gave each group.

My roommate and I both wanted to make cookies last night. She wanted to make sugar cookies, and I wanted to make chocolate chip cookies. We did not have enough of the ingredients for full batches of both our recipes, however, and the grocery stores nearby were already closed for the evening.	We have 3 eggs total. Chocolate chip cookies need $2\frac{1}{4}$ cups of flour. Sugar cookies need 1 cup of brown sugar.
We decided to split up the ingredients and make partial batches of each kind of cookie. Show how you can split up the ingredients we have to share. You do *not* have to use all the ingredients!	Chocolate chip cookies need 1.25 cups of brown sugar. Sugar cookies need 2 cups of white sugar. We have $3\frac{3}{4}$ cups of flour total.
Sugar cookies need 2 eggs. We have $1\frac{5}{4}$ cups of brown sugar. Chocolate chip cookies need 1 cup of white sugar.	Chocolate chip cookies need 2 eggs. Sugar cookies need $1\frac{1}{4}$ cups of flour. We have 2 cups of white sugar total.

Fig. 5.4. Clue cards for It's Cooking Time

IT'S COOKING TIME TASK CARD

MATERIALS:

Markers

Chart Paper

Clues and Envelopes

PART 1

1. Resource Manager gathers materials for the group.

2. The Team Captain will read, or choose someone to read, the "It's Cooking Time" task card before beginning the task.

3. The Facilitator will then open the envelope and randomly disperse the clues to all the team members. (Depending on the team's size, some team members will need to read more than one clue.)

4. *Only the person who is handed the clue is allowed to read the clue.*

5. One at a time, read a clue and work together to solve the problem.

6. One clue card has the problem you are trying to solve.

7. Use the information found on the other clue cards to solve the problem.

8. Each team member is responsible for writing down notes in their journals detailing how the group solved the problem.

9. You have twenty minutes to read through the task, gather your materials, find your question, and formulate the answer.

PART 2

1. Create a poster that includes the following:

   a. The question

   b. The answer

   c. The process for finding the solution

2. Make sure the poster is readable, neat, and labeled clearly and correctly.

3. I will choose which team member will present the information, so make sure your entire team is prepared to present the information to the whole class.

4. You have ten minutes to complete your poster.

Fig. 5.5. Task card for "It's Cooking Time"

We would describe the task card in figure 5.5 as highly scaffolded: it tells students in some detail what to do and how to do it. This level of scaffolding makes good sense for students relatively new to complex instruction, who have not yet fully learned the roles and norms. As children become more skilled at doing math this way, we suggest less scaffolding, leaving more organization to the group. As you become more expert at using the roles and norms and holding students accountable for these responsibilities, you will not need to include all this scaffolding on the task card. This will give your students opportunities to do more problem solving and get more experience figuring things out together—that is, figuring out not only challenging math, but also how to help and make sense of one another's ideas. Figure 5.6 is a much simpler version of the task card that uses the same clue cards.

It's Cooking Time Task Card

Using the clue cards, your team is responsible for figuring out what fraction of a batch of each kind of cookies your group can make with the ingredients given. Each person should solve the problem on his or her own paper.

Final Product: Your team must create a final product that includes the problem, your team's solution, and an explanation of how your team solved the problem. Each person must be ready to present the information.

Fig. 5.6

Your decision about exactly how much scaffolding to use will depend on many things, including how much experience you and your students have had with complex instruction and how well your students work in groups. Some teachers will decide to use less scaffolding than Teresa but more than we have built into the second task card. You might also decide to include on the task card the information given in the two top left clue cards and use only the other four cards as clues.

We offer a similar caveat about Helen's "Discussing Area" task card: that card is also highly scaffolded and should be simplified for teachers and students accustomed to working with roles and norms.

Some teachers design multiple abilities treatments almost every time they teach a group-worthy lesson, whereas others use them much more sparingly. Teresa did not use one to introduce her lesson, but what she wrote about the task above helped us design one she might have used. Reasons for using one might include getting students to think harder about what skills group members might bring to the task or what "doing math" involves beyond adding and subtracting. A multiple abilities treatment for this task might look like figure 5.7.

In order to succeed at this task, your group will need to do the following:

- Read carefully
- Interpret clues
- Convert fractions to decimals and decimals to fractions
- Imagine a cooking scenario in order to organize information
- Design graphics to communicate information
- Convert between improper fractions and whole numbers

In order to succeed at this task, your group will need to do the following: *Continued*

- Keep track of ingredients for two different recipes
- Use logic
- Use elimination
- Compare amounts

No one has all these abilities, but together, the members of your group have the abilities necessary to succeed at this task.

Fig. 5.7. A multiple abilities treatment for It's Cooking Time

We try to mix up the computational and noncomputational skills when we post a multiple abilities treatment, and we try to start and end with a noncomputational skill. These two moves reassure students who lack confidence in their computational ability that they will, in fact, have skills to offer their groups. Mixing up the computational and problem-solving skills also says that we consider all these skills important to doing math.

We present the multiple abilities treatment to our students at the start of the lesson, in order to expand their ideas of what doing math involves and to sensitize them to the range of intellectual strengths their classmates may bring to the work. Cohen's research (1997) indicates that the more a teacher uses multiple abilities treatments, the more low-status students participate in the group's work. Her research also shows that as teachers become more skilled at using the multiple abilities treatment, the treatment becomes more effective in equalizing participation.

By itself, however, the multiple abilities treatment is unlikely to accomplish much: we need to take further steps, even before we launch the lesson, to ensure that all students participate in the group's work. We have laid the groundwork by teaching our students new norms for groupwork, helping them learn the roles that will support more equitable participation, holding them accountable for using the norms and roles, and assigning them a challenging mathematical task that will require a range of intellectual strengths and offers entry points to all the children. We can improve the likelihood of equal participation—and learning—by attending to accountability structures as we design our lesson.

Building in Individual and Group Accountability

Almost everyone who has spent much time working in groups remembers occasions in which one or two people did almost all the work and others did next to nothing. Many teachers give up on groupwork because they have seen this happen, and they find the differences in students' engagement and learning troubling. In order to avoid this kind of unequal participation, we must hold students accountable for *both* their own learning and that of their groupmates.

As teachers, we are typically quite good at individual accountability: no matter how they teach, most teachers evaluate students' work frequently. When we ask students to work in groups, we need to continue our focus on individual accountability while also creating a means of requiring individual students to take on responsibility for the group's work, including the learning of other students. By attending to their groupmates in this way, students enhance their own learning. They must make sense of, assess, and respond to others' ideas. They must present, defend, and modify their own ideas. Thus the intellectual work of moving beyond individual accountability moves children toward deeper understanding of the math they are studying.

Being responsible for the work of others requires students to learn a new set of skills. They must learn how to ask questions that elicit others' ideas, how to listen and make sense of what others say, how to convey their own ideas so that classmates understand, and how to check to see if everyone has similar understandings. Because this is new, different, and more challenging work than merely recording one's own thoughts, we must specifically teach students how to do these things. In addition, we must also hold students responsible for attending to others' learning. If we do not make students accountable to the group, and for everyone's learning, students will find ways to accomplish the group's physical work without engaging in the intellectual work. For example, each group member may be given work to do, but students will not actually attend to everyone's ideas. For that reason, they will not learn as much as they might if they worked to make sense of ideas that they did not at first understand.

In order to build structures into lessons that include both group and individual accountability, we suggest that as you plan a groupworthy task, you consider including the following four elements:

1. An emphasis on one or two specific norms and skills for eliciting, listening to, and sharing ideas

2. A group product with clear, explicit mathematical criteria

3. An individual product that requires understanding the focal mathematical ideas

4. A plan for checking in with individuals to assess their understanding of the group's work

Although this is a short list, this is also a lot to incorporate into any one lesson. So, as you are starting out with complex instruction, you might focus specifically on one or two of these elements and then add more of them as you feel ready to do so.

1. An Emphasis on Specific Norms and Skills for Eliciting, Listening to, and Sharing Ideas

As we describe in chapter 3, when you teach groupwork norms to your students, you will have to make the norms concrete by teaching specific strategies for productive interactions.

For example, as you focus on the norm of helping others, you will teach students to how to help one another—not to tell answers, but to offer explanations. Although your students might become quite adept at these norms in lessons that specifically focus on them, as they engage in rigorous mathematics, mathematical challenges may distract them from recently learned norms, leading them to revert to more comfortable but less productive ways of working together. In order to thwart the gravitational pull of old habits, your lesson objectives should include one or two specific groupwork norms and accompanying skills that you want to re-emphasize.

The list of norms in chapter 3 may give you ideas about which ones to address during a particular lesson. You should plan to talk explicitly about these norms at the start of the lesson; you should also plan how and when to give students feedback about their performance relative to the norm. For example, as we plan a lesson, we write the norms we want students to think about on our task cards. Then, we might plan to record on the board or the overhead examples of things we heard students say that show how they are enacting the norms. Finally, our plan would include a closing discussion about students' performance on the norms. If you are looking for ideas for reinforcing norms during a lesson, you might browse chapter 6, which contains several specific strategies for addressing status and encouraging equitable participation.

2. A Group Product with Clear, Explicit Mathematical Criteria

The directions you write for the final group product provide one important tool for focusing students on a task's important mathematical ideas. You'll want to think carefully about how to design a group product that requires all students to demonstrate their mathematical thinking, and then you'll want to explain your criteria for the group product clearly to your students. For example, if we were using the "It's Cooking Time" task, we might decide that groups should produce a poster that shows their solution. However, for some students, producing a poster evokes a desire to design and decorate. As a consequence, vague directions frequently lead to posters with nonmathematical features such as flowers, pretty borders, or cartoon characters. Although these posters are attractive, they actually contain very little math, and producing the poster fails to engage students in mathematical conversations. Instead, the group product for "It's Cooking Time" has students articulate the problem, and explain their solution. For students new to groupworthy tasks, we would probably craft more explicit directions, perhaps stating on the task card that everything they put on their poster should have a mathematical purpose that contributes to explaining the problem and solution. We could then plan to walk around as students are working, asking groups about various aspects of their poster and how those features are mathematical.

Giving students clear criteria for evaluating group projects pays off richly, both in the quality of group products and in individual learning. The research on this matter is unambiguous: when students know how the teacher will evaluate the group product, they have more mathematical discussions while working, they produce a better group product, and they learn more (Cohen et al. 2002). Moreover, when students write individually about the math they did in the group, as we often require older elementary and middle school students to do, those who have clearer criteria for the group product write better individual essays. The effect of explicit directions on individual learning appears to be *through* its effect on the group's conversation and the quality of the group product. This point is enormously significant: research does not tell educators as much as we would like to know about how to set the stage for conversations that will result in group learning, but this research provides one strategy for improving groupwork and individual learning.

One way that giving groups clear evaluation criteria improves the discourse in groups is by reducing the amount of off-task talk. Studying complex instruction classrooms, Cohen and her colleagues (2002) found that groups that did not receive evaluation criteria were off task 19.5 percent of the time, whereas those who did receive criteria were off task only 12.7 percent of the time.

Constructing useful criteria takes thought, practice, and, ideally, opportunities to compare notes with colleagues committed to the same enterprise. As we noted above, the standards we design for evaluating projects should reflect the task's academic content and point to connections between the activity and the central mathematical concept. They should also require multiple abilities. Moreover, if the criteria are to improve group products and group discourse, instructors must teach students how to use them and must actually employ the criteria when giving groups feedback. Indeed, the criteria are important partly because they give teachers guidance in making their feedback more specific (Cohen et al. 2002).

3. An Individual Product That Requires an Understanding of the Task's Primary Mathematical Ideas

Frequently, when we design groupwork tasks, we focus on the group product and fail to hold students individually accountable for the lesson's mathematical ideas. Students also tend to direct their efforts toward completing the group product and forget to make sure that they understand the task's content. In order to keep everyone focused on learning, we recommend that you include some means of individual accountability. We frequently require students to make notes as the group works. We might also end a lesson by requiring students to write individually about what they have learned. You could also construct a group task but require individuals to generate their own product at the end.

For example, one group of kindergarten and first-grade teachers added individual accountability to a cooperative logic task taken from the Lawrence Hall of Science materials (see appendix A). In this logic task, children used a series of clues to figure out which one of six different bears was the Mystery Bear (see Goodman and Kopp 1992, pp. 44–45). The groups received six bear picture cards, each of which showed a bear wearing different clothes, showing different emotions, and holding different objects; and four clue cards, each of which said something about the Mystery Bear's clothing, facial expression, or possessions. Because the task as given in the book did not address individual accountability, these teachers added this piece by giving all students a blank outline of a bear and asking them to draw the Mystery Bear after they and their group had discovered which one it was. This addition made students accountable both as a group, when they needed to work together to read the clues and see which pictures each clue excluded, and as individuals, because each child had to pay attention to the group discussion in order to know what features to include on his or her drawing. Having each student draw the bear added another dimension to the task, affording even more opportunities for students' engagement.

4. A Plan for Checking In with Individuals to Assess Their Understanding of the Group's Work

In addition to requiring an individual product, you might consider whether you will check in with individual students at any point during the lesson. Checking in during a lesson has two purposes: it can be an intervention for status problems, and it can provide a means of assessing students' learning.

If you are worried about potential status problems with particular students and groups, you might plan to watch those groups carefully. This, however, can be tricky because hovering over a group will undermine its autonomy and distract the children from probing one another's ideas about the mathematics. Try instead to stay in the background, checking in only when you see inequitable participation.

You might also plan to check in with students once their group has completed a task. In the example in chapter 3, Joe Cleary held individuals and groups accountable when he required each student in each group to explain a part of the group's solution before allowing anyone in the group to move to the next problem. Students needed not only to show him the group product with each step completed, but also to demonstrate understanding of what the group had done.

Holding students accountable both for their own learning and for that of their groupmates can disrupt status relationships, because all group members have a stake in their own learning, in that of all other group members, and in the quality of the group product. Over time, as children observe all their classmates contributing in important ways to these goals, their perceptions of one another's competence will change. Moreover, holding students accountable for their mathematical understandings during groupwork will create rich opportunities for teachers to learn more about how their students are thinking, which new ideas they are beginning to understand, and which ones confuse them. This kind of ongoing assessment is essential to good mathematics teaching.

This chapter has described how multiple abilities treatments and accountability can address status issues by providing students information about what they will need, individually and as a group, to succeed in a task. In chapter 6, we will describe ways a teacher can intervene *during the lesson* when she sees status issues disrupting some students' participation and learning.

References

Cohen, Elizabeth G. *Designing Groupwork: Strategies for the Heterogeneous Classroom.* 2nd ed. New York: Teachers College Press, 1994.

————. "Understanding Status Problems: Sources and Consequences." In *Working for Equity in Heterogeneous Classrooms: Sociological Theory in Practice,* edited by Elizabeth G. Cohen and Rachel A. Lotan, pp. 61–76. New York: Teachers College Press, 1997.

Cohen, Elizabeth G., Rachel A. Lotan, Percy L. Abram, Beth A. Scarlos, and Susan E. Schultz. "Can Groups Learn?" *Teachers College Record* 104, no. 6 (March 2002): 1045–68.

Goodman, Jan M., and Jaine Kopp. *Group Solutions: Cooperative Logic Activities for Grades K–4.* Berkeley, Calif.: Lawrence Hall of Science, 1992.

Stein, Mary Kay. "Mathematical Argumentation: Putting Umph into Classroom Discussions." *Mathematics Teaching in the Middle School* 7, no. 2 (October 2001): 110–12.

6

Addressing Status Issues during the Lesson

S O FAR, we have explained how teachers can prepare their students for groupwork by teaching new norms and roles (chapter 3), identified points they should keep in mind as they choose or design a mathematical task for groups (chapter 4), and described steps they can take as they plan a math lesson to increase and equalize participation (chapter 5). Once the teacher launches the lesson, he or she must perform a bit of a balancing act. The research on complex instruction shows that the more teachers intervene directly in groups, the less the children talk to one another about their academic task (Cohen 1997; Cohen, Lotan, and Holthuis 1997). Since the conversation among students is what leads to learning, teachers must, to the greatest extent possible, delegate the authority for making sense of the mathematics and completing the task to the groups. When issues of status seem to be undermining a group's work and the children's learning, however, teachers will want to intercede. Complex instruction provides some pedagogical strategies for addressing status issues that arise *during* the lesson. Some of these are *assigning competence, calling huddles,* and *giving participation quizzes.* You can learn more about assigning competence from the work of Elizabeth Cohen (1994). Lisa Jilk and her colleagues at Railside High School developed huddles and participation quizzes as strategies for addressing status issues that arise during a math lesson, but they have not yet written about their work on this.

Assigning Competence

Teachers have enormous power—often, more power than they believe they have when they try vainly to keep their students at work on the task at hand—to enforce norms of civility and mutual respect, or to help children understand the sense behind the actions taken to subtract 28 from 300. *Assigning competence* is a way to use this power (1) to help high-status students appreciate the intellectual contributions of lower-status classmates, (2) to help low-status students appreciate their own "math smarts," and (3) to call attention to useful strategies that one or more students have devised in working on a problem. As the examples below demonstrate, assigning competence can help all students, but perhaps especially those who have done poorly in math in the past, see that real math presents a broader array of intellectual challenges than they had realized.

As we have said before, students to whom others have assigned low status often say very little during groupwork. Their classmates do not expect much from them, and they do not expect much of themselves. When they do offer suggestions, they may speak in a quiet, tentative way that virtually assures that others will ignore their idea. Sometimes it seems that the students with high status barely see the lower-status students they are supposedly working with. But like high-status students, low-status students have talents, skills, and ideas to contribute to the group, and the vigilant teacher can often catch a child with low status offering an idea that no one in her group notices. When we see this happening, we can accomplish quite a lot by drawing the group's attention to this otherwise invisible contribution. In the introduction, we saw Elise Murray find a somewhat indirect way to assign competence to a low-status fourth grader. You might want to return to that story now and notice the way Elise, the teacher, got the high-status boys to pay attention to their lower-status groupmate's idea.

Notice that Elise allows the group to work through the clues on their own, giving them intellectual authority for their group's process, and waits until they believe they have finished the logic task before approaching them. By addressing her questions to the entire group and then listening hard, Elise located competence that the three boys had totally missed. She drew the boys' attention to Annette's idea simply by locating herself as far as possible from the low-status child so that the girl would have to speak more loudly than usual in order for her teacher to hear her. By drawing the boys' attention to this new, useful idea, Elise raised Annette's status, at least temporarily. Having a high-status boy spontaneously label an idea "so smart" was an excellent way to "assign competence," and thus improve a student's status.

Assigning competence is the practice of drawing public attention to a given student's intellectual contribution to a group's problem-solving efforts, as Elise did during the Build-It task. Like Annette, Miguel, the Spanish-speaking third grader we introduced at the beginning of chapter 2, had skills that his classmates had not noticed. When the other third graders looked at him, they saw a poorly dressed troublemaker who spoke little English. Aware of Miguel's intellectual strengths, his teacher, had been waiting for a chance to make his skills visible to the other third graders. When she saw him using the diagram on a task card to build a sturdy straw structure, she moved quickly (Shulman, Lotan, and Whitcomb 1998, p. 70).

> I knew that this was the chance I had been looking for; it was clear that Miguel had the ability to build things by following diagrams. I decided to intervene, speaking both Spanish and English, since not everyone in the group spoke Spanish. I told the group that Miguel understood the task very well and would be an important resource because he had a great ability to construct something by looking at the diagram, I also said that Miguel might grow up to be an architect, since building structures by following diagrams is one of the things architects need to do. I also told the group they had to rely on their translator so Miguel could explain what he was doing.
>
> I continued observing the group from a distance and, sure enough, a few minutes later the translator was asking Miguel for help. Miguel explained to the members of his group what he had done and why. It was obvious that he had abilities that could help him succeed in cooperative learning groups, and his group finally realized it. But I wanted everyone in the classroom to know that Miguel was very good at building structures. So when his group reported on their work, I said that I had noticed that the group had had

some problems understanding the task, and I asked the group reporter what had helped the group complete the task successfully.

The reporter told the class that Miguel had understood what to do and had explained it to the group. I then reinforced the reporter's explanation, adding that Miguel had shown competence in building things by looking at a diagram and that his contribution had helped his group solve the problem successfully. By assigning competence to Miguel in front of his group and the whole class, I made sure everyone knew that Miguel had a lot to contribute to his peers. This was a wonderful example for everyone of how important it is [to] explore the multiple abilities of all group members in completing the task. After this, things changed for Miguel. His fellow group members not only recognized him as an active member, but began using him as a resource to help them balance their structures.

It is important to notice that both Elise Murray and Miguel's teacher drew attention to a low-status student's *intellectual contribution*, not to a neatly made diagram or a colorful poster. Their message to the other students: "This classmate can help you with the *thinking* that your assignment requires. He knows things that you have not yet figured out."

Elise made sure that Annette's group heard and appreciated what Annette had to say. Miguel's teacher went even further: she brought Miguel's skill to the attention of the entire class, labeled it, and explained how it had helped his group solve the problem. By assigning Miguel competence, Miguel's teacher changed his classmates' perception of him in a way that made a long-term difference in his participation and added to the intellectual resources available to the other children in the class.

Assigning competence can serve multiple purposes. It can raise a low-status child's standing in the class. It can also help expand *all* the children's ideas about what doing math involves and introduce them to new mathematical strategies. We see the latter in Annette's case, where the boys learn that solving a logic puzzle requires attending to several different conditions simultaneously instead of plunging ahead with the step-by-step approach that works well with arithmetic problems. Third, assigning competence supports the norm of group accountability. The teacher insists that the other students actually need to do more than acknowledge the new strategy's validity: they have to struggle to understand it so as to use it in the group product. And, finally, assigning competence demonstrates to the class that *all* their classmates are intellectual resources for their learning.

We might be tempted to see assigning competence as something we should do privately so as not to embarrass low-status students by putting them in the spotlight. Or, we may believe that a lack of engagement is a problem with an individual student, and so we should address it on an individual basis. Status issues, however, are *not* an individual student's problem; they are a group's problem and, most of the time, the problem of an entire class. Rather than assume that a disengaged student is lazy, unmotivated, or unwilling to collaborate, we should carefully watch the actions of everyone in the group. Frequently, observing the group this closely produces evidence that other children are systematically excluding the apparently disengaged student. She may have tried to contribute an idea, only to have it dismissed or ignored, or she may have tried to ask questions, only to have other group members rebuff them. When we locate the blame for lack of engagement or excessive engagement with one student, we fail to see how each group member's actions have shaped the participation of others. If we only address the low-status student's non-participation, we don't change other group members' beliefs or actions, which renders our efforts to resolve the status issues ineffective. A problem in a group or class is rarely just a problem with one individual.

Opportunities to assign competence to a low-status child are not, of course, available all the time. Wanting to show his classmates that Miguel had more skills than they recognized, his teacher kept a vigilant eye on him for months, waiting for her opportunity.

"I realized the only way to change students' views about Miguel was to show them that he had certain abilities to contribute to his group," his teacher writes. "My challenge was to identify his strengths and show his peers that he was competent" (Shulman, Lotan, and Whitcomb 1998, pp. 69–70). It was spring before she saw her chance. Elise, however, found an opportunity to assign competence to a low-status student—and saw positive results—the first time she taught a complex instruction lesson.

In addition to its effect on the way classmates see a child, assigning competence can have important consequences for low-status students' perceptions of themselves. Some students with low status have very little confidence in their own mathematical skills and abilities. Unlike Miguel, who was eager to participate and confident in helping his group, students who believe they have nothing to contribute may prefer to hide behind the protection of nonparticipation conferred by their low status.

Marcy Wood worked with such a student during a summer math remediation class for eighth graders. This student, Alonzo, contributed only reluctantly to group conversations and rarely asked for help when stuck on a problem. In fact, Alonzo had a reputation as a troublemaker, a difficult student eager to disrupt his classmates' learning. This changed after Marcy successfully assigned competence to him.

Marcy's eighth-grade students were trying to find the area of an oval shape (see fig. 6.1). The oval was drawn on a square grid so students could find the area by counting squares. One challenge was figuring out how to count squares that were only partially covered by the oval. Some students tallied every square that was inside, or partly inside, the oval, counting those that the oval only half covered as full squares; the resulting answer was too big. Others counted only squares that were completely within the oval, ignoring any partial squares; this gave an answer that was too small. Only Alonzo developed a strategy for measuring the area that accounted for the partial squares. He imagined putting the parts together to make whole squares and then counted these whole squares along with the squares that the oval completely enclosed. This was a significant mathematical move that demonstrated Alonzo's understanding of area, nonstandard measurement, and relationships among parts and wholes.

After visiting individually with students about this problem, Marcy decided the children could benefit by discussing one another's strategies. She got students who had counted every square and students who had counted only whole squares to present their work to the group. Alonzo's work was the key to pushing students to new understandings of area, but he was reluctant to present. His classmates teased him and insisted that he was not the smartest student in the class. Marcy persisted, nonetheless, and eventually persuaded Alonzo to present his ideas. He quietly explained his solution to the class. A fair amount of debate ensued, ultimately persuading the other students that Alonzo did, indeed, have the best strategy for finding the area.

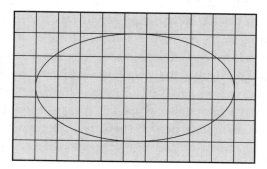

Fig. 6.1. An area problem

From this moment, Alonzo was a different student. He no longer disrupted the classroom, but instead worked quietly on his own. Although still reluctant to ask for help, he accepted assistance that Marcy offered and willingly presented his work to the class. He even volunteered to stay in during his break to finish work on a problem. Although we cannot prove that assigning competence, by itself, transformed Alonzo's behavior, the evidence suggests that public recognition of his intellectual accomplishment played some role in persuading him both that he was smart in math and that engaging with math problems was more worthwhile than derailing his learning and that of others.

Assigning competence will not always, of course, change children's behavior dramatically. However, it is a powerful tool for focusing students, and ourselves as teachers, on students' intellectual strengths, and it is a primary resource for implementing complex instruction in our classrooms. It is also an important tool for giving *all* students access to new mathematical ideas. When we insist, as Marcy did, that other students acknowledge a new idea and that they struggle with it until they understand it, we force them to stretch intellectually, see a piece of mathematics in a new way, and connect a novel approach to one that seems more obvious to them. This process deepens and extends mathematical learning. When a teacher makes his students accountable for understanding their groupmates' approaches to a mathematical challenge in this way, he ups the lesson's mathematical challenge and the potential intellectual payoff.

We want to emphasize here that teachers need to assign competence to students who have high and average status as well as those who have low status. One reason for this is obvious: children will notice, as we want them to, when you assign competence. If the low-status students' ideas and achievements are the only ones that the teacher mentions publicly, assigning competence may begin to have an effect very different from the one the teacher desires: it may, after a while, perversely ratify a child's low status. Obviously we want to avoid that outcome. However, there is another good reason for assigning competence to many different children: assigning competence directs students' attention to a good idea or helpful practice. Low-status children are not the only ones who do or say things we want others to notice and think about. Also, assigning competence can be a powerful tool for helping all students identify additional mathematical smartnesses. If Charlene has high status because of computational facility and we can assign competence to her for another mathematical skill, we help Charlene and her classmates value, and then use, other mathematical abilities.

In order for teachers to assign competence usefully, they need to think about both the strengths individuals bring to the math class and their own learning objectives for the lesson. As a teacher learns more about her students, she can purposefully construct tasks that require the smartnesses of particular students who have low status with their classmates and strengths that high-status students have yet to develop. Similarly, teachers' expanded understanding of mathematics as a discipline and mathematical methods helps them identify and name students' competencies in the moment. This is really hard to do right now, because the current testing frenzy pushes districts and schools to embrace a very narrow vision of school mathematics.

In current work in secondary school math classrooms, Lisa Jilk and teachers in and around Seattle have found it valuable to assign competence to individuals and groups for moves that contribute to a group's smooth working as well as to those that advance work on the mathematics. In this way, Lisa says, teachers can expand their students' ideas about what it means to be a math learner as they deal with status concerns; they can also hold students accountable for the way they function in a group. A teacher might say, "I really like the way Team 4 is checking in with one another on problem #2. That's going to help you out a lot with the rest of the task." Or "Beth, you do a great job of moving your work to the center of the table when you talk. That helps everyone in your team see what you're working on." Other processes a teacher might assign competence to are asking good questions, explaining ideas, checking with a team member, helping without giving answers, and pressing for understanding—in short, any of the norms she is trying to teach.

Huddles

Sometimes, after you launch a task, you realize that things are not proceeding as you had hoped and expected. Perhaps hands are waving all over the room, indicating that students do not understand the task card. Or perhaps you observe that, in two or three groups, one student seems to have taken over, or some groups are ignoring a group member. Huddles, an idea developed by math teachers at Railside High School, are a strategy for addressing these difficulties.

To call a huddle, a teacher announces to the entire class that he wants all children who share a particular role (all recorders, for example, or all materials managers) to meet in the back of the classroom. Once this huddle has assembled, the teacher can clarify directions, give hints, or provide needed information. If used strategically, huddles can also work to enhance the status of particular students. If you notice that a group is excluding one of its members, you might call all students sharing his role to huddle with you. Giving important information to these students may allow a teacher to help an excluded student to find a way into his group's conversation, without singling him out publicly.

In a math lesson we observed, Lisa Jilk, noting that the reporter in one group had had few opportunities to contribute to her group's discussion, called a huddle, announcing to the whole class, "I need to see all the reporters over here, please!" When she had all of them huddled around her, she gave each one a transparency sheet and the instructions they needed to create an overhead that would display their group's ideas to the other workshop participants later in the hour. Lisa watched the student she was worried about return to her group and confidently deliver the information about the final task. Throughout the rest of the lesson, this student found more opportunities to make her voice heard. In this instance, the huddle seemed to help a student find her footing in a group and opened a door to increased participation.

It is important, however, that this information be truly useful. Ferrying unneeded directions from the teacher will not enhance a child's status in the group. It can actually undermine a student's participation, too, if the teacher pulls her out at a bad time and she can't find a way to re-enter the group's conversation. In addition to ensuring that students in the huddle bring back something vital to their group's work, the teacher must coach them on how to present this information to their group so that the important ideas get communicated. Otherwise, the group huddle is not effective in reducing status issues, and it could even perpetuate them. For example, one of us observed a low-status elementary school student return from a huddle with information his group needed and then struggle in vain to explain the teacher's directions. Finally he shrugged, saying that he guessed the group should continue with what they were doing. A group member rolled his eyes and said, "We already know that!"

In order to make huddles more effective, one teacher we know, Susan, who you will learn more about in chapter 8, makes a point of summarizing the significant information at the end of the huddle, then asking the students she had called to the huddle to tell her what they will say when they return to their groups. If she needs to, she then restates the important points.

Susan also uses huddles to disrupt status in her classroom by assigning competence and enforcing norms. For example, during one mathematics lesson, Susan called a huddle of facilitators. When they were gathered around her, she stated that she had noticed that some groups had simply divided the task into small pieces, so that individuals were solving their individual pieces and then simply copying one another's answers. Then she asked one of the facilitators to describe what was happening in his group. He said that they were talking as a group about each part of the task. Susan asked him some specific questions about how this was working and what he was doing to help this happen, assigning him competence as an expert at facilitating group discussion. She then suggested some phrases that the other facilitators could use to get their groups to discuss the tasks instead of simply sharing answers. Before the facilitators returned to their groups, she suggested, "Let's remind one another what is it that you're going to say when you return to your group." After several facilitators had suggested language for helping their groups talk together about the math, Susan dismissed them to their groups. The student to whom Susan had assigned competence returned to his group, briefly interrupted the group, which had continued working in his absence, and articulately restated the groupwork expectations discussed in the huddle.

Huddles permit teachers to intervene in groups without disrupting the work flow. Because only one group member leaves for the huddle, work on the mathematics task can continue. Thus an announcement about the work or a management issue does not interrupt the whole class. For this reason, using huddles supports the growth of a classroom culture where engagement with mathematical problems is the norm. In addition, huddles offer a way of ascribing status to the student who goes to the huddle and then returns to the group with vital information. And, as Susan demonstrated, you can also use huddles to meet instructional needs, such as reminding students about norms and roles.

Participation Quizzes

A *participation quiz*, another idea from Railside High School, offers an additional way for a teacher to hold children accountable for enacting the norms and roles they have learned. While students work in groups on the mathematics, teachers can take notes on how they work together, their use of language from the task cards, their enactment of classroom norms, and individuals' intellectual contributions. Taking these notes publicly—on an overhead, under a document camera, on a whiteboard, or on chart paper—is one way of communicating to children which behaviors their teacher values and of encouraging actions that minimize status differences. By writing down productive comments she overhears and good working strategies she observes, a teacher can give groups language and ideas about how to work effectively as a team. A teacher can also use the participation quiz to emphasize particular norms or to help students develop subtler skills, like helping one another without giving answers. Sometimes groups ignore this public commentary. Sometimes, however, students pay close attention to the comments a teacher is writing and adjust their behaviors as they work. We have overheard, "Hey, I said that!" as a student notices a sentence on the board. Debriefing about these notes at the end of the math period can help students understand what it means to enact a norm in a particular situation.

In order to provide feedback to individual groups, a teacher might divide a sheet on the overhead or under the document camera into five sections, one for each of her five groups. This teacher might make notes like those in figure 6.2 as the class works on a math problem.

These notes serve two important functions. They give students feedback on what they are saying that helps their group, *and* they offer all students examples of what they might say to help move their groups' work forward. The notes should address two aspects of groups' work: the mathematical moves that are helping students make progress on the problem and the social moves that are helping all students participate.

In addition to helping a teacher work on norms and roles early in the year, participation quizzes can be useful later—perhaps in the dark days of February—when children seem to have forgotten some of the norms and are working together less productively. The participation quiz can remind them of better ways of doing and learning math. Teachers Lisa Jilk has worked with have used participation quizzes when they changed groups, when they gave a group test, when students just seemed "off" and needed to be reminded of how they should be acting. The participation quizzes are also great for the beginning of the school year, when students may need a bit of an incentive, or sometimes a somewhat higher-stakes intervention, to learn how to collaborate.

Group 1

"Wait, not everyone is finished."

"Let's take turns trying to explain our answer because we know that Ms. K will ask us, 'Why?'"

Group 2

"I think that we need to figure out which ingredients go in which cookies."

"I think we need to convert this mixed number into a fraction."

Group 3

"I don't get it—yet."

"5/4 of a cup must be 1 and a fourth because four fourths is one, and then there is another fourth."

Group 4

"Miguel, can you read your clue again? I think we forgot part of it."

Group 5

"Listen, guys! Melissa has a good strategy. Melissa, say it again."

"Is anyone besides me confused?"

"Let's make a chart."

"Let's read our clues again to see if they all work."

Fig. 6.2. Notes from a participation quiz

One teacher devised a way of recording group participation using diagrams with boxes and arrows. For example, figure 6.3 indicates that the task's materials were in the middle of the table, and that all group members had access to them and were participating in the task's work. In comparison, figure 6.4 shows that the four students in the group were working in pairs.

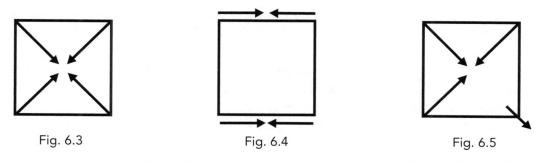

Fig. 6.3   Fig. 6.4   Fig. 6.5

Figure 6.5 shows us that three group members are working together and one is working alone or is elsewhere.

Once students become used to this public note taking and have talked a few times about what they see on the overhead, you can use notes of this sort as a "quiz" that assesses students, formally or informally, on their groupwork skills. In these situations, a teacher can let children know before the lesson the kinds of behaviors he will be looking for and can keep open-ended notes or develop a checklist focused on specific behaviors he wishes to see.

Focus areas for participation quizzes could include the following.

1. Discussing mathematical ideas
2. Using language on the role cards
3. Talking only within the group
4 Encouraging all members' participation
5. Completing all tasks
6. Calling the teacher over only for group questions
7. Using ideas from all group members in the final product

Each behavior above will increase the likelihood that all students will participate productively in the task. By noting these behaviors as part of a quiz, you can see—and make visible to your students—the progress they make over time. This will help you appreciate progress and also tune up problem areas.

Occasionally a teacher may choose to include more critical comments on the participation quiz. This could happen if one group persistently divides up the intellectual work despite the teacher's earnest efforts to get them to talk together about their ideas, or if a few individuals seem to be sabotaging others' efforts. We would not advise such a public use of negative comments as a first-line strategy for addressing concerns, but teachers should certainly consider it if other approaches have failed.

While the Students Are Working Together

This chapter focuses on moves you can make during the math lesson that address status issues. We want to emphasize at this point that, in addition to keeping track of status and participation, the teacher should attend to the children's work on the mathematics. We use complex instruction to teach math precisely because we think that groups, when they function well, create an excellent setting for learning. Learning is an active process, and in groups, children must try out their ideas, justify them, and respond to others who are trying to make sense of these ideas and offering their own thoughts. Individual seatwork often fails to demand the same kind of active intellectual involvement. But precisely because groups furnish a setting for powerful learning, we need to be sure that the math that children learn is what we want them to learn. Moreover, unless we listen in on groups with an ear for the mathematical ideas that children explain and investigate, we lose out on one of the primary advantages groupwork offers the teacher: when children explain their thinking out loud, they offer us an unrivaled opportunity to learn about their ideas, both insights and misconceptions. This kind of formative assessment of students' understanding allows us to teach math much more effectively. The more we learn about our students' thinking, the better we can tailor our instruction to their actual mathematical needs. Here are some suggestions for you as you enact your lesson.

Launch the Task

State both the mathematical and the groupwork objectives. We frequently specify a lesson's mathematical objective. However, if we want students to focus on working well as a group, we need also to state groupwork goals explicitly at the start of a lesson. Here is how one teacher set up a task that involved negative numbers.

> Today we're going to be working with ideas about adding and subtracting negative numbers. We've been focused on this for a few days, but we still need to get better at working with these numbers. Understanding this been really challenging for a lot of us.
>
> We've also been working on asking each other good questions when we do our math together. In order to do this task, you'll need to be asking good questions so everyone really understands why you tried something or how you made sense of it…. As I walk around the classroom today, I want to hear you asking each other questions about how you solved the problem, or why you did something.

You have to be clear about what you expect if you want students to participate in particular ways, and if you intend to hold the children accountable for their participation.

As Your Students Get to Work

Observe the conversations in the groups with an eye on the mathematics. What mathematical work are the students engaging in? What ideas and skills are they bringing to the task? What ideas are they struggling with or misunderstanding? Are they learning what you hoped they would? Observing is a bit tricky, because you do not want to hover: the children need freedom to think about the problem together, without an adult sailing in with suggestions. Still, if they are heading in the wrong direction, but seem serenely oblivious to their plight, you want to know. Often, in this situation, we hear ourselves asking the off-course group, "How's it going?" They, more often than not, cheerily respond, "Great!" and send us on our way. More specific questions like, "Tell me what your group is thinking right now," "What has this group figured out so far?" or "How did you start?" create more openings for questions that zero in on their confusion or give you information about how they are thinking mathematically.

Let's imagine, for example, that your students are working on It's Cooking Time (see chapter 5), and that one group has decided to change all the mixed numbers into fractions by adding numerators and denominators (see fig. 6.6).

$$2\frac{1}{2} \text{ cups of flour} = \frac{2}{1} + \frac{1}{2} = \frac{(2+1)}{(1+2)} = \frac{3}{3} = 1$$

Fig. 6.6

You could listen to the group's conversation from a distance in hopes of hearing some-one quietly voicing doubts, perhaps wondering whether $2^1/_2$ could really be the same as 1. If you did hear such a contribution, this would be a good moment for assigning competence, moving in toward that group, and remarking, "That's an important idea, Raphael. Think-ing about whether your answer is reasonable will help your group solve this task." Or, you might decide that these four children are far too pleased with their labor-saving strategy for converting fractions to pay attention to anyone who suggests that the answers they got do not make sense. In that instance, you may ask, "What has this group figured out so far?" This kind of question leaves you an opening for asking other questions that will challenge them to rethink, whereas "How's it going?" creates no openings at all unless the students ask for your input.

Toward the End of Group Time

You will need to decide how to challenge and extend the work of groups who finish early. Some teachers put an extension at the bottom of the task card: "If you have time, ...". You might, for example, tell groups who have figured out one way to make both kinds of cookies in the It's Cooking Time task to come up with a second solution that gives them more sugar cookies than the first one, but still allows them to make part of a batch of chocolate chip cookies. If you prefer to keep the task card simple, you could keep an eye on the groups and tell any that finish before others to see if they can find a second solution that will result in less leftover flour.

Wrap Up the Lesson

As the math period draws to a close, you should also be thinking about what you want to do to close the lesson. What big mathematical ideas did some groups discover that you should make visible to everyone else? Do you want to bring up with the whole class anything about how the children worked together? Do you want to reflect with the children on the math-ematical and groupwork goals that you mentioned when you launched the task?

During the wrap-up, you will want to draw attention to a few important ideas emerging from the children's work. Although many children, and perhaps every child, will want to present their thinking, you should not let this happen. With so many presenters, the group will probably lose focus. Children may also come to think that they do not need to pay atten-tion during the wrap-up time if they understood what their group did. If the latter happens, you have lost a good venue for teaching both mathematics and norms for groupwork.

Summary

In chapters 3, 5, and 6 we looked at pedagogical strategies that teachers can use to address status problems. In chapter 3, we looked at how assigning children specific roles and creating new norms for groupwork can help equalize participation. We also considered some strategies that teachers use to teach these roles and norms. In chapter 5, we looked at what a teacher can do when planning a lesson to increase the chances of equitable participation. Here in chapter 6, we considered teaching moves the teacher can make *during* the math lesson.

But as valuable as the roles, norms, and strategies are, the character of the mathematical tasks that groups work on can be just as important in altering participation and status perceptions in math lessons. After teaching and analyzing his first complex instruction lesson, one student teacher, Sean, reflected on the mathematical demands of the task his group had given the students and on its impact on the students he was most concerned about. He and his classmates had originally thought the task was mathematically complex and thus groupworthy. Reflecting on the lesson later, however, Sean concluded that the task's complexity was really all about speed and facility with difficult computations; it did not require or reward other mathematical smartnesses. He had tried to equalize status in groups by assigning coveted roles to low-status students. However, because of the task's emphasis on computation—the exact skill the high-status students possessed in abundance and the low-status students lacked—the lesson only reinforced existing status differences. Sean writes,

> Because I felt that our task was not ... a true groupworthy task, I do not feel like the roles really affected the participation of the low-status students as much as we hoped. I felt that having the low-status students be in charge of the materials and the fake money helped the students to be a little more involved, but it did not allow the students to show the group their "smartness" and therefore did not raise the students' status.

As Sean notes, if we want children to see and appreciate what their quieter or low-status classmates have to offer, we must give groups mathematical tasks that require a variety of "math smarts." So, we turn our attention now back to the design of groupworthy tasks.

References

Cohen, Elizabeth G. "Understanding Status Problems: Sources and Consequences." In *Working for Equity in Heterogeneous Classrooms: Sociological Theory in Practice,* edited by Elizabeth G. Cohen and Rachel A. Lotan, pp. 61–76. New York: Teachers College Press, 1997.

Cohen, Elizabeth G., Rachel A. Lotan, and Nicole C. Holthuis. "Organizing the Classroom for Learning." In *Working for Equity in Heterogeneous Classrooms: Sociological Theory in Practice,* edited by Elizabeth G. Cohen and Rachel A. Lotan, pp. 31–43, New York: Teachers College Press, 1997.

Shulman, Judith H., Rachel A. Lotan, and Jennifer A. Whitcomb, eds. *Groupwork in Diverse Classrooms: A Casebook for Educators.* New York: Teachers College Press, 1998.

Selecting and Designing Groupworthy Tasks

WE HAVE talked at length about strategies for making small groups places where children will all participate and learn mathematics. As important as these strategies are, children will not learn mathematics unless these groups have challenging, mathematically rich, groupworthy tasks to work on. In chapter 4, we looked carefully at one task, showing how the task's features support teachers' efforts to involve all children with the big ideas that form the foundation of third-grade math. We now turn our attention to finding and creating tasks that will support your efforts to move children toward a deeper understanding of the math they need to learn in your class.

You can get some help in finding good tasks in a number of sources that may be familiar to you. The National Council of Teachers of Mathematics (NCTM) *Standards* documents (1989, 1991, 1995, 2000) have mathematically meaty tasks that you can adapt easily to complex instruction groupwork. Marilyn Burns's *About Teaching Mathematics: A K–8 Resource* (2000) also has some excellent tasks for groups: you can adapt many of these for different grade levels without compromising their educational potential. We think, for example, the "$1.00 word" problem (Burns 2000, p. 16), displayed below, is both groupworthy and mathematically challenging. You could use it as is in grades 3, 4, or 5.

1. If a = $.01, b = $.02, c = $.03 and so on, what is the value of your first name?
2. Using this alphabetic system, one of the days of the week is worth exactly $1.00. Which one is it?
3. Find other words that are worth exactly $1.00.

We also highly recommend *Elementary and Middle School Mathematics* (Van de Walle, Karp, and Bay-Williams 2010), not only because it offers mathematically meaty problems, but also because it explains clearly what is difficult in each curriculum area. Prospective teachers often find it so useful that they do not sell it back to bookstores, so used copies can be hard to find. The book has been through many editions, however. We believe that *all* editions are useful, and the earlier ones are somewhat cheaper. Figure 7.1 gives an example of the tasks the book offers.

	If this rectangle is *one-third* of the whole, what could the whole look like?
	If this rectangle is *three-fourths* of the whole, what could the whole look like?
	If this rectangle is *four-thirds* of the whole, what could the whole look like?

Fig. 7.1. (VAN DE WALLE, JOHN A.; KARP, KAREN S.; BAY-WILLIAMS, JENNIFER M., ELEMENTARY AND MIDDLE SCHOOL MATHEMATICS: TEACHING DEVELOPMENTALLY, 7th Edition, © 2010, p. 298. Reprinted by permission of Pearson Education, Inc., Upper Saddle River, NJ)

Appendix A will say more about resources you can use to find good tasks.

We know, however, that you almost certainly have a mathematics curriculum to follow, as well as particular mathematical skills and knowledge that you need to teach. Building on what we have said in earlier chapters, especially chapter 4, about characteristics of groupworthy tasks, this chapter offers you guidance in adapting and designing tasks that will engage students, challenge them, and support their learning of significant mathematics, specifically the mathematical ideas and skills that your curriculum emphasizes.

In either looking for a task in the math materials you already have or designing a task from scratch, you will start with the mathematics you want your students to learn, the big ideas that you want them to engage with. Figuring out the big ideas may take a bit of work. Textbooks tend to organize around operations and skills rather than ideas. Conversations with one or two other teachers will probably help. Even if your colleagues teach different grades and do not use complex instruction, if they have in the past helped you think about your students' mathematical confusions, they can probably help you identify the big ideas that undergird the math you teach. Van de Walle, Karp, and Bay-Williams (2010) is another good resource for this. Identifying the big ideas is only the first step, but it is a crucial one.

We begin our discussion of task construction with an example from a fifth-grade teacher we admire. The task requires students to examine what they are doing mathematically when they use the standard multiplication algorithm. It engages them with place value as it manifests itself in multiplication, with the meaning of partial products in this computation, and with relationships of the numbers with one another. This initial section of the chapter presents the task itself, mathematical goals and expectations that the teacher had when he designed it, and some students' work. We then describe several ways that we have adapted the task for use with other learners. Finally, we explain why we like the task and what we have learned from working with it.

The chapter's second section lays out criteria that guide us in creating and revising groupworthy tasks and offers guidance for those wishing to create their own tasks. We know that most teachers need to ground their math curriculum in state and district standards, which may connect to the Common Core Standards for Mathematics that many states have adopted, and in the textbook that their district has chosen. Section three offers suggestions for locating good tasks in your textbook and adapting ones that have some good features but are not groupworthy as written. Most, although certainly not all, textbooks *do* contain at least a few tasks that will work well for groups with only minor alterations. In section three, we present three problems that we found in textbooks and recommend almost as is, explain why we like them, and offer task cards that a teacher might use when assigning them to groups.

As promised, we start with a fifth-grade task and explore how small adjustments to the task can significantly increase students' interactions with important mathematical ideas.

A Fifth-Grade Task: 15 × 49

Joe Cleary, a fifth-grade teacher in Holt, Michigan, wanted to deepen his students' understanding of the meaning of multiplication. During the first weeks of his unit on multiplication, he used a number of tasks that involved his students with different representations and solution methods for a variety of multiplication problems.

For the end of the unit, Joe developed a groupworthy task that he used before students took the unit test. He had the following goals.

■ Help students bring together various ideas about multiplication

■ Help students summarize and synthesize what they learned

■ Help students "show off" their new understandings and realize how much they had learned

Concerning this task's origin, Joe writes (Cleary, personal communication, August 30, 2010):

> As for the 15 × 49 task, it evolved from an Investigations in Number, Data, and Space (TERC 2006) lesson on multiplication. The lesson in Investigations asked the students to think about multiplication as clusters of problems and to work on solving equations as a series of "clusters" or partial products. I had worked with my class on these kinds of clusters and on other ways of solving multiplication equations. It was this work that led me to create a complex task that asked students to bring together their strategies.

Figure 7.2 gives the original version of Joe's task. Before you read any further, try this task yourself.

Joe chose the numbers 15 and 49 because of the many partitioning and compensation strategies students could use to solve this problem. For example, he predicted that his fifth-graders might multiply 15 × 50 to get 750 and then subtract 15, because one fewer 15 would give them 49 × 15 rather than 50 × 15, to get a final solution of 735, as shown in figure 7.3.

As a group, construct a poster that shows four ways of solving and explaining the following number sentence:

$$15 \times 49 = ?$$

Your explanation might include grids, pictures, charts, algorithms, written explanations, or any other way your group can think of to explain the equation.

Each member in the group should be prepared to show and explain your group's answers.

Fig. 7.2

$$49 \times 15 = ?$$
$$(49 + 1) \times 15 = 50 \times 15 = 750$$

But this is 50 fifteens and we only need 49 fifteens, so we have to subtract a fifteen:

$$750 - 15 = 735$$

Fig. 7.3

Or, they might multiply 49×10, get 490, then multiply 49×5; or take half of 490, since 5×49 would be half of 10×49; get 245; and add those products to obtain 735, as in figure 7.4.

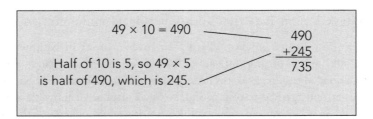

Fig. 7.4

The Students' Solutions

The fifth graders came up with a number of different ways to solve the problem Joe had given them. Every group included the traditional algorithm shown in figure 7.5.

$$^4 49$$
$$\times\ 15$$
$$\overline{245}$$
$$\underline{490}$$
$$735$$

Fig. 7.5

As Joe had anticipated (see fig. 7.3), two groups, noticing that 49 is one less than 50, computed 50 × 15 and explained, "We rounded 49 to 50 and subtracted 15 from the product."

Several groups created a grid that enabled them to compute each of the partial products separately and then to add them up, as shown in figure 7.6.

×	40	9
10	400	90
5	200	45
	600	135 = 735

Fig. 7.6

Two groups partitioned 49 into 40 + 9, computed the products of 40 × 15 and 9 × 15 using the standard algorithm, and added the two products. Several groups also added the number 49 fifteen times in a column. All the groups also had responses that were representations rather than solution methods. For example, four groups wrote a story problem that illustrated the *meaning* of the number sentence 15 × 49 = 735. Examples included the following:

Clara baked some batches of cookies. There were 15 cookies in a batch …

Tom went to the store and bought 49 packages of cups. Each package has 15 cups …

Another group included prime factorization (see fig. 7.7) as a representation of the problem.

(5 x 3) x (7 x 7)

V V

15 x 49

Fig. 7.7

Two groups represented the multiplication with an area model, constructing rectangles that were 49 units by 15 units, as shown in figure 7.8.

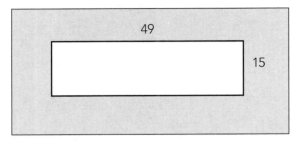

Fig. 7.8

Having asked for solution methods, Joe was surprised to see area-model representations (fig. 7.8) that identified one possible meaning for the number sentence but did not show a way to compute an answer to the problem. However, he encouraged the students to show how they might use their rectangles to solve the problem. To do so, one group divided the rectangle into 100-unit and 10-unit sections, as shown below:

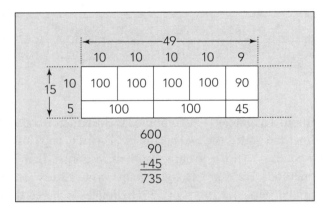

Fig. 7.9

When Joe showed us the fifth-graders' work, we saw that the students' responses, although not what their teacher had originally intended, demonstrated an understanding of multiplication's meaning in many ways. Joe reported that he was able to conclude the work on the task with a discussion in which students looked across the different groups' responses to make some comparisons and contrasts among them. The success of this last pedagogical move made us wonder whether the effort of trying to make explicit connections among various solution strategies might not cause students to dig deeper into the meaning of the numbers in the familiar multiplication algorithm. Together, we decided to alter the task by adding a directive to look for connections among their solution strategies. Figure 7.10 shows our new task.

As a group, use four different strategies to solve 24 × 15.
Your final product should do the following.

1. Show each strategy, and how you used the strategy to solve 24 × 15. You can use grids, pictures, charts, written explanations, or any other tools you need to make your explanation clear and easy to understand.
2. Show connections across the strategies.

Fig. 7.10

Again, we suggest you try this new version of the task before reading further. This time, focus on trying to identify and show the connections among your methods: for example, you might notice the numbers that occur in partial products in several of them. If you have trouble doing this, you are in good company: read on.

Prospective Teachers Work on 24 × 15

Several of us gave this new version of the task to prospective teachers in our methods classes. We describe their work here, both because it illustrates some of the strategies we expect younger students to try and because it enables us to delve deeper into the mathematical possibilities arising from our modification. Like Joe's students, the prospective teachers gave the traditional algorithm a prominent place on their posters. One class had just read a chapter that laid out many strategies for solving multiplication problems, and, as we had expected, they turned to the book to find examples of alternative algorithms. They also included a number of other algorithms that used the distributive property or compensation strategies. Several groups used one or more of the approaches shown in figure 7.11. The directive to "show connections among the strategies" puzzled the prospective teachers, however.

Solution 1	Solution 2	Solution 3	Solution 4
Traditional Algorithm	20 × 10 = 200	20 × 15 = 300	25 × 15 = 375
	20 × 5 = 100	4 × 15 = + 60	375 − 15 = 360
	4 × 10 = 40	300 + 60 = 360	
²24	4 × 5 = + 20		
× 15	360		
120			
24			
360			

Fig. 7.11

To help them see connections, we asked the groups to choose a number that was visible in the process of one solution and see if they could find it in another solution. For example, we pointed out the 200 in solution 2 (see fig. 7.11 above) and asked, "Can you find 200 in any of the other solution methods? Does it mean the same thing? What else can you find in more than one method?" We hope that the prospective teachers will someday ask *their* students these kinds of questions, and we encourage you to try these questions when using this task.

Several groups looked for the 200 in the traditional algorithm and realized that it was there in the 24, which is really 240 or 200 + 40, as shown in figure 7.11, solution 2. All groups found some numbers that showed up across solution methods. We were pleased that, by the end of the lesson, many students had concluded that all the various algorithms for multiplication involved "taking numbers apart and putting them back together." This insight is central to understanding any approach to multidigit multiplication or, for that matter, any other operation that involves place value, one of the big ideas that students work on throughout elementary school.

When we took this second version of the task—the one requiring students to look for connections among multiplication strategies—to other fifth-grade classrooms, we found that the children, like the prospective teachers, found multiple ways to solve the multiplication problem. But the suggestion that they locate "connections" among solution strategies, however, puzzled them just as much as it had puzzled the prospective teachers. Like the prospective teachers, the fifth graders needed some help thinking about what this would mean. We suspect that students might find connections more easily if we engaged them in some preliminary work finding connections across solution strategies with smaller numbers such as 12×6.

We like this task for a couple of reasons. First, it is straightforward: We ask students to think about a variety of ways to solve a multiplication problem, explain each of their approaches, and locate connections among the various solution methods. The task and the task cards that we used remind us that once students are used to using complex instruction guidelines, a task card need not be elaborate to serve its purpose adequately.

Second, nothing in this problem distracts the children from the mathematics. Often the most attractive math problems given in textbooks or other curricula involve students in making decisions that are not really mathematical. Recall, for example, the ubiquitous problems involving calculating how many ice-cream-flavor-and-cone combinations one can make with chocolate, vanilla, and chocolate-chip ice cream and three different kinds of cones. Or, take a look at the problem that we analyze in appendix B, in which students must decide where they can paint a mural and what it should look like, given particular amounts of different colors of paint. Joe's problem asks students to think first about different ways to solve and represent a multidigit multiplication problem. Our extension has them think about the connections among their different approaches. This investigation should take students directly into the mathematics.

A third reason we like this task is that it offers the students a chance to pull together what they know about place value and multiplication and showcase their understandings.

This opportunity for reflection helps students move beyond just doing mathematics to actually thinking about what they are doing as they do mathematics. As they learn how to think about their mathematical work rather than just about the mathematics, they will gain tools that will support deeper engagement in more challenging mathematical work.

Perhaps this task's most significant strength, however, is a feature that we originally saw as a weakness: none of the groups who originally undertook the task were clear on what making connections among the various strategies would mean. Even when we explained it, both children and adults needed support in identifying connections. We hesitated to recommend the task's second version to others, because we were afraid that teachers would feel discouraged if their students failed to find connections. We then realized, however, that the very fact that students of all ages had to struggle to see mathematical connections among strategies indicated that the task would challenge students who had always succeeded in solving school math problems before as well as those with a more mixed history. The second version would create an opportunity for *everyone* to learn important math with and from one another.

By asking for four different strategies, the task provides entry for children with widely different skills: even those who can't yet multiply two-digit numbers can add a column of fifteen 49s. By asking for connections, the task challenges all students, including those who often move through math assignments at warp speed. Moreover, the groups' struggles point them toward a deeper understanding of the mathematics. We want a complex instruction task to do this: it is worth the time it takes precisely because it challenges all students and pushes them intellectually, moving them toward clearer understanding of big mathematical ideas. Watching prospective teachers struggling to see connections among different ways of solving a multiplication problem alerted us to a weakness in our teaching as well as in the prospective teachers' earlier schooling: when we introduce students to new ways to represent and solve multiplication problems, we usually hope to help them grasp the mathematics behind the algorithm they have learned to use competently. If they cannot see connections between that algorithm and other approaches to a multiplication problem, we know that we have not achieved this goal. We think that the version of this task that requires students to look for connections among approaches serves the same diagnostic function in fifth grade that it does in the college classroom.

Struggling to make connections, as the prospective teachers found, helps children and adults get deeper into the mathematics. As we saw above, just about everyone will probably use the traditional algorithm to solve a multidigit multiplication problem. Making connections helps them make sense of what they are doing when they use that algorithm.

This task reminded us that it is okay if a task overshoots what students can currently do unaided. Teachers need to ask themselves whether their students, in groups, can work on it productively and whether it will engage them in supportive conversations with each other about rigorous math. Lev Vygotsky (1978) and his zone of proximal development (ZPD) remind us of the importance of presenting students with tasks beyond what they can solve independently. Vygotsky's ZPD is the space in which a learner can succeed only with others' aid. Learning occurs as students work in this space and become able to perform tasks independently for which they earlier needed help. Ideally, groupworthy tasks fall into this zone: no one in the group can do these tasks alone, but students can make progress on them if they work together and help one another. A good task affords multiple paths of entry, based on children's current skills and knowledge; pushes students beyond what they already know and can do alone; and supports them in doing together what they cannot yet do independently. When we ask our students to make connections among different strategies for solving multiplication problems, we provide that push.

Before leaving this task behind, we want to mention that teachers we know have adapted it to meet different mathematical goals, using different numbers and, occasionally, different operations. Substituting decimal numbers for 24 and 15, and leaving the rest of the task the same, will get students thinking about place value in a very different, and less familiar context. A teacher of younger children asked her students to show three ways of *adding* 49 and 15. And one of us asked students to show four different ways to compare $6/9$ and $5/8$. A group of teachers or prospective teachers could undoubtedly find many other ways to rewrite this task to serve the goals of their curricula.

What to Think about When Adapting a Task to Make It Groupworthy

Although group tasks in textbooks often have considerable potential as prompts for mathematical conversation, they can usually benefit from some editing that aims to center the children's attention on the mathematics, as Joe's task does. We ask ourselves the questions below when adapting a textbook task. Only rarely do we solve all the problems that these questions suggest. However, we do find that considering each question at least briefly is helpful. After a little practice, you will find that you can do this quickly. We have found, too, that when we work with a colleague, we can answer more of the questions to our satisfaction. Our questions focus on three different aspects of task creation: the mathematical ideas, the context the problem creates, and its groupworthiness.

1. Questions that help you to figure out what mathematical reasoning children will need to do to complete the task

A. Think first about the task's mathematical demands. Figure out what important mathematical ideas you want your students to be working on. Ask yourself first what ideas the task *could* get your students thinking about, and then which ones you want to focus on.

B. Ask yourself whether this task will challenge *all* your students mathematically. If not, figure out how you can adapt it so that it will.

C. How can you adapt the task so that it affords more than one starting point, requires more than one solution path, and thus makes some mathematical connections visible? What strategies will your students use to succeed with this task? What might they try? What prior knowledge do they bring to this task that will help them succeed?

   D. What could you add to the task to help students reflect on their mathematical
      work?

2. Questions to help you to figure out how the task's context supports or distracts from the mathematical thinking you want your students to do

   A. How are your students likely to work on this task? What would they be likely to
      talk about and do? If you think that they might focus on parts of the task that do
      not involve mathematics (e.g., "I want to have sugar cones because they are my
      favorite"), try to figure out how you can adapt the task so that it will direct their
      attention and conversation to the math.

   B. Think about what decisions students will make. Will they make these decisions
      individually or collectively? Will these decisions require thinking about math? If
      not, figure out how you can adapt the task so that the thinking and talking will
      involve math.

3. Questions that help you to figure out how you can make the task more groupworthy

   A. Does the task allow for multiple entry points and multiple solution strategies? If
      not, how can you adapt it so that it does? Will students display multiple math-
      ematical competencies?

   B. Students often approach a group task by dividing up the work in some way that
      makes sense to them (e.g., "I'll do the first two problems. Juan can do the next
      two...."). When they do this, they are less likely to talk about the math, and they
      don't get access to the learning that each of the different problems affords. If you
      think they would be inclined to divide up the work, will this division help all stu-
      dents engage in the task's important mathematics, or will it mean that the they
      finish quickly but miss an opportunity to explore important mathematical ideas?
      Think about how you can structure the task so that everyone in each group must
      contribute and do mathematical reasoning.

   C. How does the task provide for individual and group accountability? How will
      you communicate how you will evaluate both individual and group work and
      learning?

Applying These Ideas to Problems and Activities You Find in Your Textbook

Your math textbook, especially the teacher's edition, is a great place to begin looking for
tasks that fit your curricular goals and are appropriate to the grade level you teach. Almost
all textbooks offer some tasks that you can use as is, although the first few times you adapt a
task from the textbook for groups in your classroom, you will probably find it helpful to do

so with the questions just listed in mind. Often, only one or the two of those questions guide our revision of a textbook's task.

Here are three problems we have found in textbooks and the task cards we created from each one.

Creating and extending patterns

This task is part of an Everyday Math lesson, "Exploring Even and Odd Numbers, Covering Shapes, and Patterns." Figure 7.12 shows the original task as it appears in the *Everyday Mathematics: First-Grade Teacher's Lesson Guide* (Bell et al. 2002, p. 185).

Exploration C: Exploring Patterns with Pattern Blocks

SMALL-GROUP ACTIVITY

One child begins a pattern with pattern blocks. The other children in the group take turns continuing the pattern. Children also take turns starting a pattern.

A group drawing can be made of some of the patterns by tracing the blocks or by using a Pattern-Block Template.

Fig. 7.12. (Bell, Jean, Max Bell, John Bretzlauf, Amy Dillard, Robert Hartfield, Andy Issacs, James McBride, Kathleen Pitvorec, and Peter Snecker. Everyday Mathematics: First-Grade Teacher's Lesson Guide, Vol. 1. 2nd ed. © Everyday Learning Corp., 2002. Reproduced with the permission of the McGraw-Hill Companies.)

We think that this task does quite well on all three sets of questions. It requires the children to reason mathematically by collaboratively creating and analyzing patterns, which are important mathematical competencies. The context seems unlikely to distract from the math, and the task offers all children in the group multiple opportunities to create a pattern and analyze other patterns in order to extend them. It also allows individuals to create simple or complex patterns.

However, several of our task-analysis questions showed us ways to make this task more groupworthy. Reflecting on question 1D, above, we realized that none of the task's words as shown in figure 7.12 suggest to the children that they should talk about the patterns they create in order to figure out what was and was not a pattern. To address that concern, we added #3, as shown in figure 7.13, which requires students to think together about each proposed pattern and decide whether it is, in fact, a pattern. Question 3 also addresses question 2B, above, by requiring group conversation and decision making focused on what *pattern* means.

And, finally, question 3C drew our attention to the absence of any structures for individual accountability. The group drawing in the original activity took care of group accountability; we added the explaining part of the poster so that each child would need to put his or her thinking into words.

Figure 7.13 shows a task card with our variations.

Making Patterns

Your group needs:

      pattern blocks

      poster paper from your teacher

      pencil, crayons

You will be making a poster that shows four different pattern trains.

1. Person 1 makes a pattern with pattern blocks.
2. Each person takes a turn adding blocks to repeat the pattern.
3. Everyone must agree that the pattern is correct.
4. The next person makes a new pattern. Everyone adds blocks to repeat the pattern. Everyone checks to see if the pattern is correct.
5. After everyone has a turn, make a poster:

      Show your group's four pattern trains.

      Each person writes a sentence describing his/her pattern.

If you have more time:

      Show each pattern a new way. (Clap/snap, cubes, letters, words, etc.)

Fig. 7.13

As you can see, this task card gives students a great deal of guidance in distributing the task's work so that each child has a chance both to make and to analyze patterns. The card also gives each group member multiple chances to analyze other children's pattern trains to see if they are, in fact, patterns and, if they are, to add to them.

We think that teachers and children can both find this kind of scaffolding useful as they get started with complex instruction. It will help them move away from groupwork in which the high-status children dominate the group, leaving others to do either menial chores (e.g., "Denise, you should make the border around the edge of the poster") or nothing at all. As students have more experience with complex instruction and with norms and roles that emphasize equitable participation, we think you will need and want less of this kind of scaffolding. One goal of complex instruction is to give students more responsibility for learning and for one another. As you remove scaffolding from the task card, you suggest to students that they are capable of making important decisions about how to learn. As you move toward shorter, less scaffolded cards, you may at first want to review some groupwork norms before you break into groups: ask the children for ideas about how their group can ensure that everyone participates in thinking and in creating a physical product. You may also specify at the outset that the class will practice one or two norms during this lesson, for example, "talk and listen equally" (see chapter 3). You will also want to conclude work in groups with a discussion of what helped groups make mathematical progress. Figure 7.14 is a task card you might use if you think your students are ready to take on more responsibility for including everyone in the intellectual work.

Making Patterns

Your group needs:

> pattern blocks
>
> poster paper from your teacher
>
> pencil, crayons

As a group, take turns making patterns and repeating them. Each person should have a turn making a pattern and each person should have a turn adding to each pattern.

YOUR FINAL PRODUCT:

> A poster with 3 pattern trains with repeating patterns and a sentence describing each pattern
>
> One train that is *NOT* a pattern and a sentence about why *IT ISN'T ONE*

If you have more time, show each pattern in a new way.

Fig. 7.14

Mystery numbers

The activity in figure 7.15, from the teacher's edition of Scott Foresman's second-grade math book (Charles et al. 1999, p. 132A), engages the children with number combinations.

Mystery Numbers

Find the missing numbers using counters.

Materials: Paper cups with lids, 15 counters

Learning style: Kinesthetic, logical

Children work in pairs to play a game using counters in a cup.

> Demonstrate the game by first counting aloud as you place 15 counters, all the same color, in a cup. Then put the lid on.
>
> Shake the cup, spill out some counters and count them. Then ask the children, "How many are in the cup?" Write the number sentence, for example, $15 - \_\_ = 6$. List all number sentences that include 6 and 15. Point out that in each case the missing number is 9.

Children continue the game using a different number of counters each time.

Fig. 7.15. (From SCOTT FORESMAN ADDISON WESLEY MATH GRADE 2 TEACHER'S EDITION by R. Charles, C. S. Barnett, D. J. Briars, W. D. Crown, M. L. Johnson, and S. J. Leinwand. Copyright © 1999 Pearson Education Inc or its affiliates. Used by permission. All rights reserved.

You can almost use this task as is. It addresses our questions related to mathematical reasoning and context very well. We were concerned, however, about its groupworthiness. We were not sure that the children would see that more than one solution path exists. We also worried that one child would take over and do all the thinking. So we focused on the third set of questions, adding roles and directives that would, we hoped, ensure that each child would get a chance to engage with the math. In answer to question 3A, we decided that a teacher could enlarge the number of solution paths by giving each pair of children more counters than they would need to play the game. Children might then use any of a variety of strategies to figure out the mystery numbers. They could use subtraction or addition facts that they already knew, use their fingers to count on from the number of counters on the table to the number that they had just placed in the cup, or use the extra counters to model the situation by creating a second set of counters of equal size to the set they started with and either removing the number of counters they had poured out of the cup onto the table or dividing the set into its known and unknown components.

Figures 7.16 and 7.17 show two task cards. The first we created for pairs of children who are just starting out with complex instruction and may not yet expect all group members to contribute substantively to the group's thinking. The second leaves far more to the group.

Mystery Numbers

You will be making a poster that shows how you found each mystery number. Write a number sentence for each mystery number and draw a picture or write about what you did.

GETTING STARTED:

The MATERIALS MANAGER should get the materials from the front table.

DOING THE MATH:

- The REPORTER should:
  1. Put between 10 and 16 counters on the table and, with the materials manager, count the counters and agree on how many there are.
  2. Put these counters into the cup.
  3. Close and shake the cup.
  4. Then spill a few counters out onto the table.
- You should each count the counters on the table and agree on the total.
- Without looking in the cup, each of you should figure out how many counters are still the cup and write down your ideas on your own papers.
- Explain your answer to your partner. Listen to your partner's answer. On your paper write a number sentence that shows what you did.
- When you and your partner AGREE on an answer, count the counters that are still in the cup. Were you right?

Mystery Numbers—*Continued*

- On the poster paper write a number sentence that shows what you and your partner did.
- Now it is the MATERIALS MANAGER'S turn to start. Choose a new number of counters to start with and repeat the steps you took before.
- Take turns doing this until you have at least six number sentences.

Make a poster that shows your number sentences and the counters that you used to make the number sentences.

Fig. 7.16

Figure 7.17 shows the simpler card.

Mystery Numbers

Use the directions below to play the game six times. Make a poster that shows a number sentence for each mystery number, and draw a picture or write about what you did.

HOW TO PLAY MYSTERY NUMBERS

1. Pick a number between 10 and 16 and put that many counters into a cup.
2. Spill some of these counters out onto the table.
3. Without looking in the cup, figure out how many counters are still in the cup and write a number sentence that shows what you think happened.
4. Check your answer by counting the counters left in the cup. Were you right?
5. Write a number sentence that shows what you did.

Play again with a different number of counters.

Fig. 7.17

Designing a zoo

Figure 7.18 shows a task we found on the challenge page of a third-grade unit on division in the teacher's edition of a Houghton-Mifflin textbook (Greenes, Leiva, and Vogeli 2002, p. 359). These enrichment pages, included in most teachers' guides, can be a rich source of groupworthy tasks.

At the Lincoln Park Petting Zoo, there are four animal pens. One pen holds 4 llamas. One pen holds 9 goats. One pen holds 18 sheep. One pen holds 12 ducks.

Mr. Lincoln needs to show all his animal pens on the grid. Can you help him?

A pen is formed when 1 animal is placed in a box next to or directly above or below another box with the same animal. All pens formed must be a square or a rectangle, and there must be at least 1 box between each pen and the next one. Use L for llama, G for goat, S for sheep, and D for duck.

[There is grid paper on the students' workbook page for the map.]

Fig. 7.18

Figures 7.19 and 7.20 show our task and role cards, respectively. We think that the context—the animals, the opportunity to experiment with differently shaped rectangles—will help make this task attractive to third graders. We do not, however, think that the task as it appears in the book will challenge them mathematically. Although the textbook offers this problem as part of a unit on division, we note that children could—and, we think, probably would—complete the work asked for without ever thinking about division. Moreover, we find it easy to imagine one or two children taking over and doing all the work. Addressing questions 1, A–D, helped us ramp up the needed mathematical reasoning. Questions 3, A–C, helped us see how to make the task more groupworthy, to ensure that each child would have access to materials and share the intellectual work. We have increased the mathematical demand by having group members consider what they would have to do to make the smallest possible zoo (see fig. 7.19, part 2). The new task calls for spatial reasoning, work with factors, recognizing geometric shapes, and hard thinking about area. Certainly, several satisfactory answers and solution paths exist. Children who are comfortable with factors can move immediately to graph paper to design their animal pens; those still shaky on multiplication and division can use tiles or cubes to experiment with different dimensions

ZOO TASK CARD

Together, you will design a zoo by creating pens for four animals: llamas, goats, sheep, and ducks.

PART 1

1. Your role card tells you which animal pen to draw and the area of the pen. On white graph paper, draw a pen to hold your animals. It must be a rectangular array. Label the array with your animal and a multiplication sentence to describe the area.

2. Each person should show their array to the group and have group members agree their work is accurate. Group members should initial

ZOO TASK CARD—*Continued*

each other's arrays if they agree they are correct. Each person should then cut out his or her array.

3. As a group, arrange your arrays on the green graph paper to make your zoo. Glue them down when you agree.

4. Together, draw a rectangle around your entire zoo. What is its area? Label this on the paper.

PART 2

1. Discuss how you could house your animals in the zoo with the smallest possible area.

2. When you agree, draw a model of this "small" zoo. Each person should draw the array for his or her animal. All pens must still have the required number of squares, and the zoo and the pens must still be rectangular. If need be, you can change the dimensions of the pens you designed in part 1.

3. Explain why you think that you have designed the smallest zoo that will house all these animals. What did you do to make the zoo smaller?

Fig. 7.19

Guide	Time Keeper
• Finds compromises among group members. • Makes sure everyone understands the task. *"I'll call the teacher over."* *"Remember, no one is done until we are all done."* Builds an 18-square pen for the sheep.	• Makes sure each member of the group participates and understands. • Keeps track of time. *"I don't get it yet… Can someone help?"* *"We need to keep moving so we can…"* Builds a 9-square pen for the goats.
Resource Manager	**Reporter**
• Makes sure everyone has the supplies they need.	• Records or finds someone to record the information.

Resource Manager— *Continued*	Reporter—*Continued*
• Takes care of and returns supplies.	• Makes sure each member of the group has a chance to say what they know.
• Organizes cleanup.	*"How should we start our poster?"*
• Calls teacher over.	*"Has everyone had a chance to say what they think?"*
"Do we have all the supplies?"	
"We need to clean up. Can you ... while I ...?"	Builds a 12-square pen for the ducks.
Builds a 4-square pen for the llamas.	

Fig. 7.20

At this point you should think about whether, and how, you would want to change the task or role cards if you were to use them with your students.

Summary

When we search a textbook for mathematically rich tasks that we can adapt fairly readily for use by groups, we ask ourselves the following three questions.

1. How can I make sure that this task will engage my students with big mathematical ideas?

2. How can I improve the chances that the context that the task proposed supports rather than distracts from mathematical investigation?

3. Does the task allow for multiple points of access and more than one solution path?

In most textbooks, especially in the challenge problems in the teacher's edition, we find a few tasks that will answer all these questions to our satisfaction without changing the task at all. Many more problems do well on some of the questions but need work in one or two areas. The questions and subquestions that we lay out above can help us identify the areas that need work.

As we adapt the problems we find in textbooks to make them groupworthy, we need to make decisions about how much to scaffold the ways of working on the task so that everyone gets a chance to contribute ideas, listen and be listened to, and ask questions. The need for this kind of scaffolding is likely to diminish over time, as the teacher becomes more skilled with managing groupwork and working with students' mathematical ideas, and the children grow more used to the norms and roles of complex instruction groupwork.

In this chapter, we have looked at tasks that needed only modest tweaking in order to meet the criteria we have identified. Appendix B presents a task we thought children would find engaging, but which needed substantial work to produce multiple points of access and involve children with significant mathematics.

These chapters present, we realize, a daunting picture of what teaching elementary school math using complex instruction entails. We have led you through the thinking that we do as we adapt mathematical tasks and tried to give you some picture of what is involved in creating good tasks from scratch. And we have urged you to attend both to the mathematics with which the groups engage and the issues related to participation that we have discussed in earlier chapters. At this point we imagine that you are asking yourself, "With all the other things I need to attend to, can this possibly be worth the effort?" We answer this with a heartfelt "Yes!" We recall the faces and voices of the teachers and student teachers we know as they reported on complex instruction math lessons. They described children participating who have never before volunteered an idea or an answer in math class. They recounted the enthusiasm and excitement of children who had not, in the past, been challenged by problems that were suitable for the rest of the class. Their stories and their enthusiasm convince us that teaching math with complex instruction is very much worth the effort.

In chapter 8, we paint a fuller portrait of a few teachers who have made complex instruction a part of their practice.

References

Bell, Jean, Max Bell, John Bretzlauf, Amy Dillard, Robert Hartfield, Andy Issacs, James McBride, Kathleen Pitvorec, and Peter Snecker. *Everyday Mathematics: First-Grade Teacher's Lesson Guide*, Vol. 1. 2nd ed. Chicago: Everyday Learning, 2002.

Burns, Marilyn. *About Teaching Mathematics: A K–8 Resource.* 2nd ed. Sausalito, Calif.: Math Solutions Publications, 2000.

Charles, Randall I., Carne S. Barnett, Diane J. Briars, Warren D. Crown, Martin L. Johnson, and Steven J. Leinwand. *Scott Foresman-Addison Wesley Math: Grade 2—Teacher's Edition.* Menlo Park, Calif.: Scott Foresman-Addison Wesley, 1999.

Greenes, Carole, Miriam A. Leiva, and Bruce R. Vogeli. *Houghton Mifflin Mathematics Teachers Edition Grade 3*, Vol. 2. Boston: Houghton Mifflin, 2002.

National Council of Teachers of Mathematics (NCTM). *Assessment Standards for School Mathematics.* Reston, Va.: NCTM, 1995.

————. *Curriculum and Evaluation Standards for School Mathematics.* Reston, Va.: NCTM, 1989.

————. *Professional Standards for Teaching Mathematics.* Reston, Va.: NCTM, 1991.

————. *Principles and Standards for School Mathematics.* Reston, Va.: NCTM, 2000.

TERC. *Investigations in Number, Data, and Space.* Boston: Pearson, 2006.

Van de Walle, John A., Karen S. Karp, and Jennifer M. Bay-Williams. *Elementary and Middle School Mathematics: Teaching Developmentally.* 7th ed. Boston: Allyn & Bacon, 2010.

Vygotsky, Lev S. *Mind in Society: The Development of Higher Psychological Processes.* Cambridge, Mass.: Harvard University Press, 1978.

8

Tales of Three Teachers

WE HAVE learned much of what we know about complex instruction from the teachers and prospective teachers who learned about it with us and then made complex instruction part of their teaching practice. Their stories contribute to all the chapters you have read. Their ideas have shaped our ideas. Their questions have driven many of our conversations about complex instruction. And, perhaps most important, their enthusiasm for their work with complex instruction has fueled us with the energy we needed to write this book.

We want to close this book with portraits and voices of three of these teachers and of the work they did in the first year they used complex instruction. These teachers worked in districts with somewhat different student demographics and varying amounts of support and resources. They used complex instruction in different ways, at different times, and to meet slightly different goals. Inevitably, they fitted complex instruction into their teaching practices in ways that reflected both the particular pressures they faced in their schools and districts and their own, very personal concerns as teachers. But all have found teaching math using complex instruction deeply rewarding. We hope their stories will put a face on the rewards and challenges of teaching with complex instruction and help you to imagine how you might begin this work in your classroom.

Joe

Joe Cleary, a white man, taught fifth grade in an intermediate school that serves a racially and socioeconomically diverse community. The school's immediate neighborhood was a study in contrasts: cornfields abutted busy intersections, and large, newly constructed homes sat next door to older motor homes. The school itself, a low brick building surrounded by narrow parking lots, lacked a playground, so students jumped rope, played games, and talked in designated indoor areas during recess.

Teaching is Joe's second career; before becoming a teacher he was a department store manager. He had been teaching for fifteen years when he attended our workshop on using complex instruction. He signed up for the workshop because, as a collaborating teacher for student teachers from Michigan State University, he wanted to learn about the teaching strategies we would be recommending to student teachers in our math methods courses. Before attending the workshop, Joe had sometimes asked his fifth graders to work in pairs, but he had never used larger groups in his math class.

Joe worked collaboratively with his colleagues and described his school as team-oriented and collegial. The school was, in fact, particularly well organized for cooperative work. Every Wednesday, the district dismissed students early so teachers could work together on lesson planning, curriculum development, grade-level assessments, and professional development. The school organized teachers and students into teams of two or three classes, each with two or three teachers who taught different subjects. Joe taught mathematics and social studies on his team. Joe's team regularly ate lunch together in his classroom, taking turns bringing food for the group. They discussed problems involving their students; talked about upcoming events, like the library night for students and their families; and planned lessons.

Although Joe had collaborated with teacher colleagues for many years, he had never, during his years as an elementary school, secondary school, or college student, had a chance to do math as a part of a group. He recalled feeling alone, adrift, and lost in these school and college classes. However, during the complex-instruction summer workshop, he experienced a level of collaboration that he found surprisingly high, even given his teaching context. He learned mathematics content and math for teaching that was new to him, and he felt like a successful learner. The conversations helped him to make sense of mathematical ideas and deepen his understanding, both of the math that he needed for teaching and of that beyond the scope of the grade level he taught. The experience contrasted dramatically with the isolation and confusion he had felt in school and college math courses. Joe commented on the mathematical tasks he worked on during the professional development.

> There's no doubt, had I had to face those on my own, I would not have succeeded the way I did there. The collaborative aspect allowed me to be successful at the task, allowed me to learn a lot of math that week.

Joe described this experience as an epiphany. It moved him to create similar experiences for his students. He had always tried to challenge all his students. By teaching the fifth graders to work together using the complex-instruction structures that had supported his own learning at the summer workshop, Joe believed that he was able to move far closer to that goal.

He began by using groupworthy math tasks, such as the ones featured in chapter 7, as culminating activities for units, because they allowed students to articulate accumulated understandings and because this seemed like a manageable place to start.

Joe loved watching his students, animated and engaged, working together on challenging problems and taking responsibility for one another's mathematical learning. He delighted in seeing them come to define success as gaining an understanding of something individually and as a group, instead of as simply finishing a task. Both he and his students came to expect every child to succeed and understand. He thus saw real, productive collaboration among students for the first time. About this new expectation, he said:

> Everyone in your group is going to succeed. It's just going to happen. I don't think I had that notion. I think I maybe had a notion that it would be nice if everyone in the group

> succeeded. And the big argument [other teachers] always seem to have [about groupwork] …—this idea that one person or two people carry the group—that's legitimate, except that you don't look at it that way. I don't look at it that way anymore. I look at it as everyone has a responsibility to succeed and to build this collaborative sphere.

We observed a complex instruction lesson in Joe's classroom, one we mentioned in chapter 3, that illustrates this expectation nicely. Joe's assignment required the students to call the teacher over once their group had finished the task's first part, in order to get permission to move ahead. We observed one group sitting with raised hands, waiting for Joe, when suddenly a student whispered loudly, "Hey! Hey, everyone. He is going to make us all explain it to him. Let's practice before we call him over! We have to make sure we all get it!" The students quickly lowered their arms and leaned toward the center of the table. They took turns explaining the group's work to one another. When each group member had done this, with the occasional comment from the other group members to correct, encourage, or add details, they sat back and raised their hands again. When Joe arrived and asked students to explain, they did so easily. Joe saw this happen routinely. He concluded that the time and effort spent teaching norms for groupwork had led students to care about whether they and their group members could explain mathematical ideas. He said,

> It's not an easy thing to do when kids haven't done it before. But over time, they start to see. I know I experienced it [during the summer workshop], and I know that I've seen it in my classroom now…. I heard it in my classroom, "He's going to make us all explain it!" Over and over I heard kids saying, "No, we don't really have a group question. He has a question, maybe." Things about kids being responsible for themselves and each other more have been transformative for me.

We all want our students to care about their own learning and that of their classmates, but few of us have found this easy to achieve. Joe's experience shows us that students can learn important mathematics from one another and begin to be concerned about whether their classmates understand the mathematics they have worked on together. Joe's students cared in a way that went beyond making others feel good, although that probably happened, too. They cared about mathematical understanding: they had learned through experience, as Joe did in the complex instruction workshop, that caring and challenging one another mathematically happen in tandem. It takes time and work to create a culture in which this happens, but belonging to a class that values and expects both caring and rigorous mathematics is rewarding for teachers and students.

Additionally, Joe was pleased to discover that complex instruction lightened some of his logistical load, shifting more responsibility from him as the teacher to his students. The norms, which promote cooperation and autonomy, resulted in fewer questions about directions. For example, Joe taught and enforced the norm "Call the teacher for group questions only," so students had to ask one another questions about directions before summoning their teacher. The fifth graders learned this norm slowly. Early in that first year, Joe reported, "I walked away nine times for every time I answered a true group question." Nine times out of ten, he explained, the child with the raised hand was the only one who even knew what the question was. However, once the fifth graders saw that Joe was serious about answering only "group questions," very few called him over to ask, "What are we supposed to do?" Teaching students to solve their problems by asking one another questions, reading and rereading the task card, and taking responsibility for one another's success was hard work, especially early in the year. Increasingly, though, it freed Joe to focus on math content and students' sense making.

Of course, Joe also faced challenges during that first year. He reported that teaching mathematics with complex instruction "takes a lot of energy," which is why he only did six groupworthy tasks during the first year after the workshop. He found teaching norms particularly intense. His students, like those in many classrooms, entered his class without knowing how to collaborate in mathematics. Joe worked hard early in the year to teach his students what it meant to do math in groups, recognizing that they were not likely to collaborate in mathematics unless given clear, explicit expectations and opportunities to practice these new ways of being math learners.

As we have emphasized in earlier chapters, unequal participation often undercuts the potential benefits of group work. Joe had the problems commonly associated with group work in his classroom. Some students attempted to take over the intellectual work from the group, and others were willing to watch passively. Joe acknowledges,

> Every stereotype you can think about with groupwork happens. They do. It happens. [Some students] want do all the work, or they want to be the boss. Or they want to step out; they don't want to be a part of it. You name it. I'm sure you can come up with a whole litany of things. Those things, I see them all. But a really good task and really good norms, it's amazing. It's amazing.

Although the results of teaching students to do math in groups were amazing, Joe cautioned that complex instruction was not a panacea, and he assured us that participation and learning in his classroom were sometimes problematic, saying,

> Just because we're doing complex instruction doesn't mean we don't still have these really difficult cases. This isn't going to necessarily bring complete equity to the classroom. I have some really difficult kids. Honestly. I'm not making that up: really, really difficult kids.

One might be tempted to give up on groupwork when children in the classroom behave in challenging ways. When Joe presented his work at a workshop for other educators, one teacher expressed concern about trying complex instruction because she had a student with severe behavioral challenges. Joe acknowledged that he has some similar students and that they had the same challenges during complex instruction groupwork as they did at other times. However, he felt the benefits to all students in collaboration, responsibility, and rigorous mathematics made complex instruction's approach to groupwork worthwhile.

Greta

Greta McHaney-Trice, a black woman who attended the complex instruction summer workshop the second year we had offered it, had just finished her eighth year teaching fourth grade in an urban elementary school. When she began to teach there, the school served a population that was about evenly split between white students and students of color, most middle class, but some quite poor. During her tenure at the school, however, many white middle-class families left the district. By the time Greta came to the complex instruction workshop, most of the children in her class were students of color, and more were working class and working poor, reflecting the economic decline of Michigan.

Greta's path to teaching was a long one: she worked in the school district's administrative offices for years, always hoping to find a way to become a teacher. Finally, after twenty years, she arranged to go back to college, finish her degree, and take the courses she needed for certification, while continuing to work full-time for the school district. Once she had her own classroom, she continued to search out and embrace opportunities for learning. For example, during the summer she attended our complex instruction workshop, she also immersed herself in an intense, month-long professional development experience run by the National Writing Project, embracing its goals of turning teachers into writers and transforming the approach to writing in elementary school classrooms. We knew Greta because of her collaboration on other teacher-education endeavors, and she came to the complex instruction workshop at our urging.

Before the workshop, Greta had used various models of groupwork instruction in her classroom, but never complex instruction, and she had rarely used groupwork in teaching mathematics. Describing herself as more confident in teaching writing and social studies than in teaching mathematics, Greta said that she was drawn to complex instruction because it would offer her students opportunities to move beyond memorizing material for standardized tests. She also wanted support for learning more about mathematics teaching. She believed that collaboration and critical thinking benefit students in and outside school. Initially, however, the sheer volume of ideas and strategies covered in the workshop overwhelmed her, and she worried that she would forget something important if she tried using complex instruction in her classroom. She waited until late in the school year to try out these ideas. In spite of this delay, she taught three complex instruction math lessons before the year ended.

In the weeks before Greta gave her students the first groupworthy task, they had been drowning in busyness, overwhelmed by assignments related to the district's quarterly and yearly assessments. Doing a complex instruction math lesson relieved the tension that had been building in the classroom. During the lesson, the students' participation, body language, and speech all indicated to their teacher that they felt optimistic and successful. And afterward many told her that they had enjoyed the experience and found it challenging and interesting. The tasks' groupworthy nature lifted their burdens and produced emotional rewards.

> Kids were really feeling good about doing the math. It was obvious that, …[compared] to the heavy load of the week before, it was almost like a refreshing oasis for many, because we hadn't gotten a chance to do that kind of activity while trying to get a lot of things done. [I was a] taskmaster, driven and driving them. So it was really a good break for me and them. Wow. We really needed this. Sort of like taking a vacation from school. I hate to say it! That's what it kind of felt like. From the routine of what had gone down that whole busy week and whole busy day, of trying to get through so much.

Lifting the week's burdens had positive consequences in the classroom. Complex instruction's structures altered students' interactions with one another and liberated Greta to observe and appreciate these changes. Adopting the observer role (which the structures of complex instruction allowed her to do) allowed Greta to see important things happening in her classroom that were not necessarily mathematical but often contributed to students' participation in mathematics. Once she learned to worry less about whether or not the students finished each task, Greta became more aware of her students. In an interview, she reflected:

> ❑ What I'm finding is that there are a lot of rich things that are happening that aren't math. It
>   forces me to be more in tune with some of those rich things, once I breathe and let myself
> ❑ know that it's not just whether or not they finished the task.

The opportunity to learn about her students contributed to Greta's positive experience with her first complex instruction math task. She reported that watching one group—three very social boys with a history of being silly together—was particularly rewarding for her because their use of norms and roles, along with their interest in the task, reshaped their participation. Ordinarily, this group needed teachers' frequent interventions for management issues and easily got off task. Greta had assigned groups randomly and worried when the three boys ended up together. Afterward, she reported that the group worked very well. They stayed on task, engaged in the math, and did not argue. The complex instruction structures enhanced their engagement.

The children's critical thinking during complex instruction math lessons, as well as their increased engagement, also pleased Greta. She asked the fourth graders to write reflections about what they learned from one task. She was happy to see this reflection, from a student who rarely succeeded as she had during the complex instruction task: "I think the activity … challenges our brain … it teaches kids how to think, not what to think."

Additionally, Greta reported that the challenging, open-ended math tasks gave her opportunities to learn more about students' mathematical understandings. While describing students' work generated during a task, Greta declared, "This definitely shows the diversity in what each kid got out of it." She appreciated learning something new about her students. Greta reported that the open-endedness that made the tasks groupworthy enlarged her opportunities to understand the children's ideas. "It really forces me to think about it, and really look at it, without just an end part of a right or wrong, or, 'Yeah, they got it,' 'No, they don't,'" she reflected. "It's good." Groupworthy tasks with multiple right answers and solution methods offered Greta a more complete picture of her students' thinking.

Like some other teachers, Greta found it difficult to make time for groupworthy tasks amidst an increasingly prescribed curriculum. Justifying to colleagues and administrators time spent on planning and doing complex instruction tasks was a constant challenge. To address this difficulty, when parents came for the school open house, Greta displayed on her bulletin boards the work her students had generated during complex instruction. Reminding herself of the value of learning mathematics this way was even more important than showcasing the work for visitors. The display reminded her that she valued moving beyond frantic test preparation toward critical thinking and creative, communicative endeavors and taking time to understand and appreciate students' interactions and mathematical understandings.

> ☐ Justifying this when you get a call that says your (test) scores aren't ... I would love to not have to talk about it. I would love to be able to say, "I can prove to you that this is great for your kid," which is one reason I decided to go ahead and post [the students' work]. It speaks to me. It's a constant reminder that this is valid, this is good, it has more merit. They will remember doing [the groupwork math lesson] long after they remember doing,
> ☐ "OK, you divide the denominator into the numerator and you get a mixed number."

The work that she posted, which included students' written reflections on the task, reminded Greta that groupworthy tasks prompt students to think more reflectively about their math work, engage in problem solving, and make connections. It reminded her of the depth of understanding she gained about her students by examining their work and their reflections.

Susan

Susan Harvey, a white woman, taught sixth grade at a school for grades five and six. The school was in a suburban neighborhood that had a grid of streets of older ranch houses on one side and a nexus of winding roads—lined with a solid wooden fence to shield residents of larger, newer homes from the outside world—on the other. Most of Susan's students were white, but a significant minority were students of color and many were English language learners, some of them children of international students at Michigan State University. Susan had always taught in this district and had taught sixth grade for the last ten years. She came to the complex instruction workshop because, like Joe, she was a cooperating teacher for student teachers from Michigan State University. Before this workshop she had done groupwork but had not used complex instruction. After the workshop, she used complex instruction every week in her classroom. Her sixth graders often used complex instruction structures even when they were not doing groupworthy tasks.

Susan reported that she had found teaching math with complex instruction profoundly rewarding; the biggest satisfaction was seeing her low-status students engaged and participating successfully. In the spring of her first year using complex instruction, she described how this had played out in a recent lesson:

> ☐ Having all the kids feel that they're responsible is amazing. I just did [a complex instruction lesson] with fractions for the whole group. I was just stunned, when ... the kids who at the beginning of the year would never ever have explained a word of anything, they were the ones [who volunteered to present]. I didn't choose them. They just felt so much more confident themselves that they actually were the ones that were explaining everything. It wasn't the usual kids who take charge. It was kids who never had done that before. So I would say that has changed, that has been such a fabulous moment for me. Seeing how, just by the structure and by making kids realize they all are truly responsible, that you get
> ☐ everybody invested in it.

127

Susan argued that using norms, roles, and groupworthy tasks had not only helped students to do what was assigned, but had also helped them take more initiative to move beyond their assigned roles. She reported that by teaching the students how to work together, supporting them with the complex instruction structures, and expecting them to take responsibility for their learning and that of their classmates, she could get everybody invested in mathematics. For Susan, this experience was profound.

> ❑ That's been one of the big, big, huge moments in my math that's carried—it's spilled over into everything that I teach. It's just been an amazing opportunity. I think, really, seriously, off the top of my head, that this would be the most career-changing thing that I've done....
> ❑ This has been the thing that's changed me the most.

The changes that Susan described became vividly apparent to us on a visit to her classroom. One of us, Joy Oslund, thought she recognized a girl she had seen a year earlier in another classroom. Joy assumed that she was mistaken, however, because the child she remembered was too shy to speak in class, whereas the student she saw now was participating eagerly in her small group. When Joy asked about this sixth grader, another adult told her that early in the year the child had indeed been very shy, but that she now participated regularly during groupwork. The norms and structures of complex instruction had, apparently, worked well for her.

Susan was also delighted to see students in her classroom begin to support one another's learning. Because of the norm "You have the responsibility to ask for help and the responsibility to offer it," students learned to struggle through difficult problems together and to support and help one another. This support spilled over into other parts of the curriculum. Students explained mathematical concepts to one another in more depth than they ever had before:

> ❑ You've gotta talk. You've gotta talk to each other. You've gotta try to struggle through it yourselves and support one another. I think they're just incredibly supportive of one
> ❑ another and it spills over into everything they do.

For Susan, a third reward, and an important one, was that with complex instruction, planning math lessons with groupwork took less time than it had before. She had previously spent a lot of time trying to organize groups so that they would be heterogeneous, work together well, include students with different skills and different personalities, and avoid combinations of children who had trouble working together. Through the work with complex instruction, she came to expect that any student could work with any other student in the classroom for two weeks, so she began to assign groups randomly by shuffling name cards. The simple idea of assigning groups randomly instead of engineering the groups' membership, Susan recalled, "... was so liberating to me."

Also, once students learned to ask one another questions about directions and materials, she wasted far less time on these matters. And because she used complex instruction structures in *all* lessons, Susan spent less time and effort planning logistics and explaining expectations overall.

Like all teachers, Susan experienced challenges teaching mathematics using complex instruction during this first year. To begin with, stepping back and allowing students to struggle without instantly intervening was hard for her. Her students were accustomed to having questions answered quickly and problems solved for them. Walking away without answering their questions, requiring them to seek answers to from one another, made Susan—and her students—uncomfortable at first. She recounted a time when she had gone to a group to check their understanding of a mathematical idea. The students could not yet explain the idea to her. Feeling torn, she asked them to keep working and moved on. The students felt ignored at first. However, they solved the problem, and when she returned to the group they could explain their mathematical work to her.

After seeing the results of letting students work things out together, Susan concluded that she needed to worry less about making students uncomfortable. "Walking away is hard, because we just don't want anybody to be upset," she explained. "We don't want anybody to feel uncomfortable. We've got to rethink that."

Through teaching mathematics with complex instruction, Susan saw firsthand what students could figure out on their own when she delayed assistance. She came to appreciate the role of discomfort in the learning process, especially since she had seen her students support one another through the discomfort.

Susan, Joe, and Greta

Susan, Joe, and Greta teach in different school districts and at different grade levels. Their students come to school from different homes and families. They use complex instruction in somewhat different ways, and for somewhat different purposes. Yet one theme threads its way though the stories that each of them tells—the theme of pleasure. Each of them wanted to talk to us about the pleasure they got from watching their students at work in groups. Although they emphasized different things, pleasure was palpable in each story.

All three teachers emphasized the delight they felt as they saw their students take responsibility for their own learning and for that of their classmates. Joe's language was dramatic: "Things about kids being responsible for themselves and each other more have been transformative for me."

Susan spoke with similar enthusiasm about seeing her sixth graders become more involved in their classmates' learning, commenting, "I think they're just incredibly supportive of one another, and it spills over into everything they do."

In different contexts, particular insights and joys stand out. Greta found immense satisfaction in a complex instruction lesson that followed weeks of grinding preparation for the tests that determine so much about an urban school's future. She experienced the groupwork as a kind of vacation from the toil of the previous weeks, a time when children could relax and actually enjoy their academic work, relieved of the tension that had characterized the previous weeks. "It was," she recalled, "almost like a refreshing oasis for many."

Complex instruction gave Greta a break, too, a respite from a role that the constant demands the school and the district placed her in, a break from being the "taskmaster, driven and driving them." And, she asserted, posting the work the children had done during a groupworthy task reminded her regularly of what she really thought was important. Her students would recall the complex instruction task long after they forgot the mechanical steps of creating a mixed number from an improper fraction.

Susan's challenges were different: she worried about the English language learners who had in the past sat silently during groupwork, trying to keep up with what was going on in this new language, unwilling to risk contributing an idea. She glowed with satisfaction as she described a recent lesson, "a fabulous moment," and the way it dramatized for her the change in these students and their role in the classroom.

Joe loved learning mathematics through working on groupworthy tasks in the complex instruction workshop. He loved the experience of seeing himself as a different kind of math learner, and he loved the way that groupworthy tasks created a venue for success for all his students. He told us with palpable delight, "Everyone in your group is going to succeed. It's just going to happen," adding that before the complex instruction workshop, "I don't think I had that notion. I think I maybe had a notion that it would be *nice* if everyone in the group succeeded."

The pleasure that Susan, Joe, and Greta take in what complex instruction has done for their students, their classrooms, and their roles as teachers delights us. We have rejoiced in teachers' and prospective teachers' pleasure, especially the delight they have taken in the ideas and growing confidence of formerly silent, low-status children. We have enjoyed having something substantial to offer those who struggle to offer a conceptual mathematics curriculum to *all* their students, to help children who had always struggled with math learn more than the times tables and the rules for converting an improper fraction to a mixed number.

Final Thoughts

Every school and every classroom is different. In the prologue, Lisa Jilk mapped for us her own path into using complex instruction to teach math in an urban high school. In Lisa's school, almost every math teacher used complex instruction, and colleagues met every week to discuss concerns about students' learning and plan lessons that would address these concerns. Susan, Joe, and Greta, by contrast, were teaching in schools in which no other teacher was using complex instruction. Their only contact with colleagues struggling to do this came in a few evening meetings that we organized as a follow-up to the summer workshop. We are sure that crafting a practice that includes complex instruction is easier in a setting that values and supports that work, a setting in which colleagues design groupworthy math tasks together and meet regularly to attack common problems. Indeed, our belief in the power of groups has brought us to writing this book.

But as conscious as we are of the value of working on this with trusted colleagues, we hope that we have convinced you that even if you are working alone, as Greta, Joe, and Susan were, the potential rewards, both for you and for your students, of making complex instruction a part of your practice far outweigh the costs.

We know of no set path for incorporating complex instruction into a math classroom, no prescribed steps to follow, or particular ways to "do" complex instruction. We think complex instruction offers multiple entry points and access for teachers with widely varying levels of experience teaching mathematics.

We want all students to engage wholeheartedly with mathematics and develop robust understandings of its big ideas. We believe that, because you have made it this far in the book, you must also share this goal with us. If we want to achieve this goal, we have to find ways to support children's engagement in rigorous mathematics. Complex instruction offers one way to do this. We hope that this book gives you the tools you need to make complex instruction a central part of your mathematics teaching practice, and that you and your students enjoy the work you will do together as much as we have. We would love to think that it could help you to interest a few colleagues in joining this endeavor.

Resources for Groupworthy Tasks

Burns, Marilyn. *About Teaching Mathematics: A K–8 Resource*. Sausalito, Calif.: Math Solutions Publications, 2007.

> Widely used in professional development and preservice teacher preparation, this recently revised classic includes more than 240 classroom-tested activities. It is organized into five comprehensive parts: "Raising the Issues," "Instruction Activities for the Content Standards," "Teaching Arithmetic," "Mathematical Discussions," and "Questions Teachers Ask." Not all the tasks are groupworthy, but many are.

Chapin, Suzanne H., Art Johnson, and Toby Gordon. *Math Matters: Grades K-8: Understanding the Math You Teach*. Sausalito, Calif.: Math Solutions Publications, 2006.

> This important resource for teaching elementary mathematics explains key elementary topics such as whole number computation, fractions, similarity, and measurement very clearly. It digs into the big ideas associated with these topics and what makes them challenging for children. Woven into each of these chapters are syntheses of research on students' thinking and mathematical activities that invite the reader to engage more fully with what he or she is reading. The book shows how each topic fits within the larger framework of elementary and middle school mathematics.

Cohen, Elizabeth G. *Designing Groupwork: Strategies for the Heterogeneous Classroom*. 2nd ed. New York: Teachers College Press, 1994.

> This book is in all our personal libraries, and we purchased it for the teachers who came to the summer workshops at Michigan State University. We assign chapters 3, 5, and 8 to our preservice teachers when we introduce them to complex instruction. The book includes an appendix of "group skill-building" exercises that are really useful for teaching students new norms and roles of working in groups.

Erickson, Tim. *Get It Together: Math Problems for Groups, Grades 4–12*. Berkeley, Calif.: EQUALS, 1989.

> This resource, which covers all strands of mathematics, has more than 100 problems for cooperative problem-solving groups in grades 4 and up, including the Build-It tasks mentioned earlier in this volume. Six clue cards give the information needed to solve each problem.

————. *United We Solve: 116 Math Problems for Groups, Grades 5–10*. Oakland, Calif.: Eeps Media, 1997.

> This book, a sequel to *Get it Together* (Erickson 1989), continues that book's pattern of math problems designed especially for groups. The problems focus on proportional reasoning, spatial visualization, and learning to generalize from patterns. Some materials from the book are also available online at http://www.eeps.com/products/uws_stuff/unitedwesolve.html.

Goodman, Jan M. *Group Solutions: Cooperative Logic Activities for Grades K–4*. Berkeley, Calif.: Lawrence Hall of Science, 1992.

> This resource is appropriate for teaching younger students. The tasks focus more on logic puzzles than on core mathematical content. They do, however, offer a good place to start groupwork in early elementary grades. One of our collaborating teachers uses many of these tasks to support her first graders' learning to do mathematics tasks in cooperative groups.

————. *Group Solutions, Too! More Cooperative Logic Activities for Grades K–4*. Berkeley, Calif.: Lawrence Hall of Science, 1997.

> A sequel to *Group Solutions* (Goodman 1992), this book also focuses on cooperative logic activities for grades K–4. The problems are designed to improve skills such as cooperation, problem solving, using the process of elimination, reasoning, communicating, visual discrimination, sequencing, spatial visualization, recognizing shapes, rearranging, and manipulating shapes. The logic activities span geometry, number sense, patterns, discrete mathematics, statistics and probability, function, and algebra.

Lampert, Magdelene. *Teaching Problems and the Problems in Teaching*. New Haven, Conn.: Yale University Press, 2001.

> This book shares a year of teaching mathematics in a fifth-grade classroom. In addition to offering detailed analyses of her pedagogical decisions and teaching moves, Lampert includes many wonderful math problems she used in her classroom and the work her students produced. Her tasks have many of the characteristics of groupworthy tasks that we discuss in this volume's chapters 4 and 7.

Schwartz, Judah L., and Joan M. Kenney. *Tasks and Rubrics for Balanced Mathematics Assessment in Primary and Elementary Grades*. Thousand Oaks, Calif.: Corwin Press, 2008.

> The math tasks in this book, and in others in the Balanced Mathematics series, adapt easily for complex-instruction group work. The developers—from Harvard University, the University of California at Berkeley, and the Shell Center in England—have also generated a wealth of downloadable tasks, rubrics, and students' sample work, available on the Web at balancedassessment.concord.org/.

Stenmark, Jean Kerr, Virginia Thompson, and Ruth Cossey. *Family Math.* **Berkeley, Calif.: Lawrence Hall of Science, 1985.**

> This is a rich resource of tasks meant to create opportunities for conversations about math among family members. Many of the tasks adapt well for complex instruction groupwork in the elementary school classroom. They cover a wide range of topics —number and estimation, logical thinking, probability and statistics, geometry, measurement, and calculators. The book contains many stimulating games, puzzles, and projects that entice children to work on hard mathematical ideas.

Van de Walle, John A., Karen S. Karp, and Jennifer M. Bay-Williams. *Elementary and Middle School Mathematics: Teaching Developmentally.* **7th ed. New York: Allyn & Bacon, 2010.**

> We require students in our elementary mathematics methods courses to buy this book and strongly encourage them to keep it as a resource for teaching. Indeed, we know many teachers who have dog-eared copies in their classrooms. The book covers all the math topics taught in the elementary and middle school grades. Each chapter begins by discussing the big mathematical ideas that serve as a content area's foundation and then describes, in very accessible language, current research on teaching and learning this mathematics. Each chapter also includes excellent tasks that adapt easily for complex instruction. One of our favorite chapters, "Exploring What It Means to Know and Do Mathematics," invites readers do some mathematical problem solving and then reflect on their mathematical activity. This chapter also contains many terrific, potentially groupworthy tasks.

Adapting Textbook Problems That Require Substantial Work

In November 2010, as we prepared to send this manuscript off to the publisher, we became concerned about whether we had given you, our reader, enough guidance in creating group-worthy mathematics tasks. We had shown you how we use the guiding questions to adapt a textbook task that already works well as a complex instruction lesson (chapter 7), and we had suggested some resources for finding good mathematics tasks (appendix A). We had not, however, talked about the more challenging task of adapting a textbook problem that, although it has some attractive features, needs substantial work in order to be groupworthy. Not everyone will want to engage that challenge: you may have lots of tasks in your school's mathematics curriculum that you can adapt fairly easily for a complex-instruction lesson. You may find that professional development resources such as those by John Van de Walle, Marilyn Burns, and others listed in appendix A offer you many problems that work for your students with little change. If so, you will not need this appendix. If, however, you would like some guidance in taking on the challenge of doing major surgery on textbook tasks in order to make them suitable for complex instruction, read on.

We will start with a problem (fig. B.1) that we found in a Scott Foresman–Addison Wesley fourth-grade math textbook (Charles et al. 1999, pp. 440–41).

> Some cities collect unused cans of paint. They give the paint to groups that are doing community projects, such as painting murals. Decide which site you want to paint. Help plan a mural using donated paint.
>
> **Paint Collected:**
> Yellow—$1/12$ container
> Red—$1/6$ container
> Light Blue—$1/3$ container
> Dark Blue – $1/4$ container
> Green—$1/12$ container

Purple—$^1/_{12}$ container

Paint Coverage:

2 containers—720 square feet

$1^1/_2$ containers—540 square feet

1 container—360 square feet

$^1/_2$ container—180 square feet

$^1/_4$ container—90 square feet

Possible Mural Sites:

School entrance—240 square feet

Bank construction site—420 square feet

Library patio—300 square feet

Work Together

Understand

1. *What do you know?*

2. *What do you need to decide?*

3. *What information will you need in order to make a decision?*

Plan and Solve

4. *How many square feet will the paint cover?*

5. *What color do you have the most of? How much of it do you have?*

Make a Decision

6. *Decide which mural site to paint.*

7. *Decide what kind of mural to paint. For example, do you have enough green to make a mural of a forest? Where's the brown paint for a desert landscape?*

Present Your Decision

8. *Share your decision and sketch a plan for your mural. Explain how the paint available helped you to decide which site to paint and what design to use.*

Fig. B.1 (From SCOTT FORESMAN ADDISON WESLEY MATH GRADE 4 STUDENT EDITION by R. Charles, C. S. Barnett, D. J. Briars, W. D. Crown, M. L. Johnson, and S. J. Leinwand. Copyright © 1999 Pearson Education Inc. or its affiliates. Used by permission. All rights reserved)

Because the textbook writers present this problem as a group activity, and because we thought that fourth and fifth graders might find using color attractive, we decided to think about how we might use this task in a complex-instruction lesson. We noticed one thing immediately: the children would not need to do much math besides adding unit fractions in order to complete the task successfully. Also, the textbook tells students how to proceed, giving them a list of questions to answer, step by step. We thought that any decision about how a teacher would use or change the task would depend in part on how much experience the students had had working in groups and how familiar they were with the norms of complex instruction. If the children had had relatively little experience with open-ended problems and multiple solution paths, we might use this task as is. With this sort of tight scaffolding, the students would probably succeed in doing what the book suggests even if they could not figure out how much wall the yellow paint would cover.

However, if the children had had more experience working in groups and using the resources their groupmates provide, we agreed that we would want to leave more thinking to the group. Thus, we might give only the information on how much paint of each color the group has, how many square feet one can covers, and the requirements of the various sites, and then, omitting the numbered directives, let the groups decide how to proceed.

We approached revising this task with two major concerns in mind. First, we saw that we would need to decide on our goals as math teachers. Did we want the children to work on proportional reasoning, or would we be satisfied to have them work on factors and area? Second, how could we ensure that all the children got an opportunity to do mathematical reasoning? The questions laid out in chapter 7 (fig. B.2) guided our revision.

---

1. **Questions that help you figure out what mathematical reasoning children will need to do to complete the task**

   A. Think first about the task's mathematical demands. Figure out what important mathematical ideas you want your students to work on. Ask yourself what ideas the task *could* get your students thinking about, and think about on which ones you want to focus.

   B. Ask yourself whether this task will challenge *all* your students mathematically. If it won't, figure out how you can adapt it so that it will.

   C. Ask yourself how can you adapt the task so it requires more than one solution and, in doing so, makes some mathematical connections visible.

   D. Ask yourself what you could add to the task to help students reflect on their mathematical work.

Fig. B.2

The first step in adapting this task for use in a complex-instruction lesson would be to decide what mathematical ideas you want the children to explore. The problem in figure B.1 demands work with area, fractions, measurement, and connections among all of these. It also requires that students distinguish between relevant and irrelevant information—an important problem-solving skill.

We considered the possibility of actually painting a mural, as an art project linked to the math lesson, but concluded that doing so would direct the children's attention away from the mathematics. To improve the chances that students would engage with mathematical ideas, we decided that we needed to introduce several constraints. Hoping to increase the likelihood that students would discover some mathematical connections, for example, we decided to require more than one solution and perhaps have the groups create at least two different geometric shapes of each color.

We discussed several other ways to increase the task's mathematical demands, which we list below. The task cards on pages 143 and 144 do not include all these adaptions. If you decide to use this task, which of these adaptations you choose will depend on your students' skills and what you want them to think about.

- First, in order to increase the need for proportional reasoning, require students to figure out the area each color would cover, given that a full can covers 240 square feet. (By substituting 240 for 360, we hoped to simplify some of the computations.)
- Require students to use all the paint, and specify that they paint no part of the surface twice.
- Specify the site for the mural, and one dimension (e.g., 10 feet high), but not the other. Given the mural's height, figuring out its length, or vice versa, would be a job for the group.
- Specify that the mural must consist of geometric shapes. Depending on your students' ages and skills, these could be rectangles only, or rectangles and triangles.

We then turned our attention to the kinds of conversation we thought the task would elicit (fig. B.3).

---

2. Questions to help you figure out how the context of the task supports or distracts from the mathematical thinking you want your students to do

A. How are your students likely to work on this task? What would they be likely to talk about and do? If you think that they would focus on parts of the task that do not involve mathematics, try to figure out how you can adapt the task so that they *will* focus their thinking and conversation on the math.

B. Think about what decisions students will make. Will they make these decisions individually or collectively? Will these decisions require thinking about math? If not, figure out how you can adapt the task so that the thinking and talking *will* involve math.

Fig. B.3

When we found this task in the textbook, we worried that many groups of fourth graders would spend considerable time picking a site on the basis of personal feelings about banks, libraries, and schools and then trying to decide what the mural might look like. Then, after most of the class period was gone, they would start to work with the mathematics and realize that maybe they don't have enough paint for the bank construction site. Although we want groups to make decisions, we want to limit the time spent on decisions that don't involve mathematics.

In order to get the children focused on the math right from the start, we decided to specify the mural's site and have groups figure out dimensions it could have, given the amount of paint available. Requiring the group to calculate dimensions forces them to think about area. Creating an abstract design made up of rectangles of the different paint colors will also prompt discussions of area, pushing students to ask and discuss questions like "How big a rectangle can we make with $1/2$ a can of purple paint?" "If $1/2$ a can covers 120 square feet, what dimensions could our purple rectangle have?" or "What if we wanted to have two purple rectangles of different sizes?"

We focused next on several of the questions about groupworthiness (fig. B.4).

3. **Questions that help you figure out how you can make the task more groupworthy**

   A. Does the task allow for multiple entry points and multiple solution strategies? If not, how can you adapt it so it does? Are there opportunities to display multiple mathematical competences?

   B. Students often approach a group task by dividing up the work in some way that makes sense to them. When they do this, they are less likely to talk about the math. Think about how you can structure the task so everyone in each group must contribute *and* do mathematical reasoning.

   C. How does the task provide for both individual and group accountability? How will you communicate clearly how you will evaluate both individuals' and groups' work and learning?

Fig. B.4

As it appears in the textbook, the Mural Task has students make decisions about the mural's location on the basis of the total number of square feet their paint can cover and what they wanted to include in the mural, given the relative amounts of various colors of paint. We could see that this initial version offers multiple entry points for students. However, as we pointed out earlier, some of these entry points would probably distract students from the big mathematical ideas we hoped they would explore. We could also imagine that students who had mastered three-digit addition could figure out how many square feet the paint could cover if they used all of it. Once they had finished this computation, all students who could figure out the relative size of different three-digit numbers could participate in figuring out which site they could paint, which colors they had most of, and what sort of picture they could paint using these amounts of these colors.

This analysis further convinced us that it made sense to change the problem in ways that we have already suggested, so that it makes more mathematical demands and provokes more discussion of mathematical questions. We did, however, wonder about other consequences of making these changes to the Mural Task. If we made our suggested changes (i.e., *choosing the site* in order to eliminate a distraction, *requiring the group to specify dimensions of the mural and of each colored rectangle* in order to get them to think about area, and *having them create an abstract design with squares and rectangles* to push them to think further about area), would groups still have multiple points of entry and multiple ways to work on the task? After further deliberation, we all agreed that our proposed changes still offered multiple entry points for students with different math smarts. If we had students make rectangles of different paint colors, those who had a good grasp of area could experiment with different dimensions for one or more rectangles of a given color. Those who still struggled with area could use tiles—perhaps colored ones—and graph paper to figure out what size rectangles they could make with a given color. Those who needed more challenge might even try to use graph paper to figure out the area of a triangle.

To address the second question (i.e., How can I structure the task so that everyone in each group needs to contribute *and* do mathematical reasoning?), we decided that putting each child in charge of creating one or more rectangles of a particular color would ensure that each child would need to think about area in a way that contributed to the group's thinking and product.

The Task Card and the Clues

We have given you a condensed summary of our conversations about the Mural Task. They clarified our ideas about how we might make this textbook task richer mathematically and more groupworthy. Now, we offer two versions of the resulting task card, as we did in chapter 7. The first (fig. B.5) scaffolds the students' work much more than the second (fig. B.6). We also offer clue cards (fig. B.7) that work for both versions.

PAINTING A MURAL OF RECTANGLES (VERSION 1)

Materials:

Grid paper, 5 pieces
Task cards
Purple, green, red, blue markers
Colored tiles

Task:

Your school has asked your group to paint a mural of rectangles. You have clue cards that tell you how much paint you have of each of four colors. A full can of paint will cover 240 square feet.

1. As a group, decide how to represent 240 square feet on grid paper. What would a rectangle with an area of 240 square feet look like?

2. Each person takes a clue card and calculates what area—how many square feet—their paint can cover. Each person shows how much this is on grid paper, using a marker that matches his or her color. Use your ideas from question #1 above.

3. Each person presents his or her calculations and drawing to the rest of the group. Work together as a group to correct any mistakes with the calculations or the drawing. Once everyone agrees that the calculations and drawings are accurate, then everyone should sign each person's grid paper.

4. As a group:

    a. Figure out how much wall you can cover using all the paint. Find dimensions for two different rectangular murals, each of which uses all the paint.

    b. Using the grid paper, create one mural made up of rectangles that uses all the paint of each color. Remember that you can have more than one rectangle of each color. The mural should be rectangular, your paint should cover the mural area completely, and you must use all the paint.

    c. Use numbers and words to prove that you used all your paint.

Fig. B.5. Task card for Mural Task, version 1

PAINTING A MURAL OF RECTANGLES (VERSION 2)

Materials:

Grid paper, 5 pieces
Clue cards
Task cards
Colored tiles
Purple, green, red, blue markers

Task:

Your school has asked your group to paint a mural of rectangles. You have clue cards that tell you how much paint you have of each of four colors. A full can of paint will cover 240 square feet.

Final Product:

Show on graph paper the design for your mural. The mural must be rectangular, be made out of rectangles, use all the paint of each color, and not have any bare, unpainted spots or spots where paint colors overlap. Your poster should tell how you know that you have used all the paint.

Each person in your group should be able to explain how the design uses all the paint.

Figure B.6. Task card for Mural Task, version 2

Clue 1

Your group has $1/2$ can of purple paint.
How much wall can you paint purple? On graph paper, sketch out a rectangle that uses all the purple paint. Write down its dimensions.

Clue 2

Your group has $1/3$ can of blue paint.
How much wall can you paint blue? On graph paper, sketch out a rectangle that uses all your blue paint. Write down its dimensions.

Clue 3

Your group has $1/6$ can of green paint.
How much wall can you paint green? On graph paper, sketch out a rectangle that uses all your green paint. Write down its dimensions.

Clue 4

Your group has $1/4$ can of red paint.

How much wall can you paint red? On graph paper, sketch out a rectangle that uses all your red paint. Write down its dimensions.

Fig. B.7. Clue cards

Analyzing the Mural Task

We have created a groupworthy task from the Scott Foresman activity. We have also explained why we chose to make specific changes. Was all this work worthwhile? What do we think are the task's strengths and weaknesses?

Strengths

- It has more than one correct solution.

- It engages all students in similar, but not identical, mathematical work. Each child must calculate and represent the area that the group's paint will cover. The parallels among the children's tasks make it easy for children to support one another, to help without giving answers. Each child has a different amount of paint. Having to calculate from the different amounts prevents students from simply copying one another's answers. Each child must, instead, develop a unique solution for his or her paint color.

- It requires students to translate a numeric area representation—240 square feet—to a visual one.

- It creates multiple ways to engage the relationship between fractions and area. Students might arrive at an area for half a paint can by multiplying 240 square feet by $1/2$, by dividing 240 by 2, by folding the rectangle that they constructed in half in order to represent 240 square feet, or by estimating half of the 240-square-foot rectangle.

- It offers an opportunity for students to use visual skills as they work on fitting their various painted rectangles into one larger rectangle, dividing their original colored rectangles into smaller rectangles as necessary.

- Because the final mural will be larger than what a single can of paint will cover, the task requires students to think about the relationship between wholes and parts and about fractions greater than one.

- It requires students to connect area and fractions, a relationship that textbooks use frequently.

- It creates an engaging context for exploring factors and relationships among area, length, and width.

- It requires students to explain their thinking about factors and area to one another.

Weaknesses

The weaknesses we have identified are mainly contextual: they could cause trouble in some classes, but probably will not in others.

- As we have pointed out above, students with either $1/2$ or $1/4$ can of paint could figure out the area that their paint will cover by folding the 240 sq. ft. rectangle they have created on squared paper. However, if students do not know how to compute $1/6$ or $1/3$ of 240, either by dividing 3 or 6 into 240 or by multiplying the fraction by the whole number, they may not be able to find the area that their paint will cover. We would advise teachers whose students are stymied by $1/6$ or $1/3$ either to (1) encourage a paper-folding strategy, even though it will probably be less accurate with these fractions; or (2) advise children in the groups explicitly to help one another work this out.

- A mistake in the first step, representing 240 square feet, will make completing the rest of the task more difficult. In situations like this, some teachers have the groups check in with the teacher before going on to the task's second part.

Few tasks are perfect. We think that the Mural Task's strengths outweigh its weaknesses.

Adapting the Task to Other Grade Levels and Other Mathematical Purposes

Before leaving the Mural Task behind, we want to note that you can easily adapt it to other mathematical goals and grade levels by changing the numbers on the clue and task cards. If, for example, we substitute small-integer amounts—4, 3, 2, and 1 cans of paint for the fractions of cans—on the clue card and say on the task card that one can of paint covers 10 square feet of wall, we have a problem that can engage third graders in thinking about area, factors, and measurement. For students ready for more challenge with proportional reasoning, you can substitute fractions like $2/3$ and $5/12$ for the unit fractions originally on the cards. You could enhance the challenge further by providing one of the mural's dimensions and having the students figure out the other one. And, of course, you can allow students to use other geometric shapes—for example, triangles, parallelograms, and trapezoids—when that would address your curricular goals.

We have devoted time and space to the paint problem in order to let you see how we work with the guiding questions we laid out in chapter 7. As you have seen, we do not address every question each time we revise a textbook problem for a complex-instruction lesson.

We do, however, try to think about the three broad areas of concern. We ask ourselves these three questions:

1. With what mathematics will this task engage my students?
2. How could the task's context (painting a mural, designing a petting zoo, and so forth) support or distract children from digging into the mathematics?
3. Will all my students be able to engage intellectually in their groups' work on this task?

Often, as we have seen, a textbook problem catches our attention because of what it offers in one of those three areas. When that happens, our job is to adapt the task in order to strengthen it in the other two areas. In the Mural Task, we worked hard on the task's mathematical demands and on making it more groupworthy. We did not, however, strive for perfection. As we saw here, for example, some tasks we like make no formal provision for individual accountability.

Several teachers we know identify finding good problems as their biggest difficulty. Others say that, once they got the hang of finding groupworthy problems in their textbook, they could do this quite easily. Your experience will undoubtedly depend partly on your school's math curriculum.

Reference

Charles, Randall I., Carne S. Barnett, Diane J. Briars, Warren D. Crown, Martin L. Johnson, and Steven J. Leinwand. *Scott Foresman-Addison Wesley Math: Grade 4.* Menlo Park, Calif.: Scott Foresman-Addison Wesley, 1999.